SpringerBriefs in Molecular Science

For further volumes:
http://www.springer.com/series/8898

Lingxin Chen · Yunqing Wang
Xiuli Fu · Ling Chen

Novel Optical Nanoprobes for Chemical and Biological Analysis

 Springer

Lingxin Chen
Yunqing Wang
Xiuli Fu
Ling Chen
Key Laboratory of Coastal Environmental
    Processes and Ecological Remediation
Yantai Institute of Coastal Zone Research
Chinese Academy of Sciences
Yantai
China

ISSN 2191-5407   ISSN 2191-5415   (electronic)
ISBN 978-3-662-43623-3   ISBN 978-3-662-43624-0   (eBook)
DOI 10.1007/978-3-662-43624-0
Springer Heidelberg New York Dordrecht London

Library of Congress Control Number: 2014940388

Printed on acid-free paper

Springer is part of Springer Science+Business Media (www.springer.com)

Preface

The goal of this brief is to give a summary of recent advances in novel optical nanoprobes for chemical and biological analysis. The design and application of novel optical nanoprobes for chemical and biological analysis has become a new, growing area of interest in recent years. Taking advantage of the development of nanotechnology, various kinds of nanomaterials with novel optical properties have been generated, laying the foundation of optical nanoprobes. By further integrating receptors (chemical ligand, aptamer, molecular imprinting polymer, etc.), the chemical information of binding specific targets will transform into analytically useful optical variation signals. These sensors are attractive owing to their high sensitivity, high specificity, and potential for easy quantification of species in many fields of application, such as conventional chemical and biological analysis, clinical diagnosis, and intracellular system sensing, or even single molecule detection for their nanoscale size. In this brief, we will give an introduction to several kinds of talented nanomaterials such as gold/silver nanoparticles, quantum dots, upconversion nanoparticles, and graphene. Furthermore, we mainly focus on the most recent reported strategies to design sensors that apply the optical principles of nanomaterials to detect targets employing using various detection techniques including colorimetry, fluorometry, surface-enhanced Raman scattering (SERS). The challenges and future perspectives of optical nanoprobes will also be presented, such as increasing sensing performance for real environmental and clinical samples, the design and application of multifunctional nanoplatforms, and biocompatibility research.

Lingxin Chen
Yunqing Wang
Xiuli Fu
Ling Chen

Contents

Chapter 1
A Brief Introduction to Optical Nanoprobes

Abstract Nanomaterial-based novel optical nanoprobes have been extensively developed because of their high sensitivity, high specificity, and potential for easy quantification of species in chemical and biological analysis. With the development of nanotechnology, various kinds of nanomaterials with novel optical properties have been generated, laying the foundation of optical nanoprobes. By further integrating receptors (chemical ligand, aptamer, molecular imprinting polymer, etc.), the chemical information of binding specific targets will transform into analytically useful optical variation signals, employing different detection techniques including colorimetry/UV–Vis spectra, fluorometry and surface enhanced Raman scatting. In this chapter, a brief introduction of optical nanoprobes is given in terms of nanomaterials, recognition units, and optical detection.

Keywords Nanomaterials · Recognition units · Optical nanoprobes · Colorimetry · Fluorescence · Surface enhanced Raman scattering

1.1 Nanomaterials

Nanomaterial is an important unit of the optical nanoprobe. The flourish development of nanotechnology lays the foundation of optical nanoprobes. Owing to their excellent optical properties, several kinds of talented nanomaterials such as gold/silver nanoparticles (Au/Ag NPs) [1, 2], quantum dots (QDs) [3], upconversion nanoparticles [4–6], and graphene [7] have been utilized in the optical nanoprobes. In an optical assay, target-induced aggregation or surface change of nanomaterials could cause the optical properties change. According to the most recent reported strategies to design optical sensors, various nanostructures involved in the sensing system are summarized, as shown in Table 1.1. Noble metal nanomaterials with various structures have been widely used in the design of optical nanoprobes. Typically, gold spheres, rods, popcorns, and hollow gold nanospheres were mainly applied as the substrate of optical nanoprobes with

L. Chen et al., *Novel Optical Nanoprobes for Chemical and Biological Analysis*, SpringerBriefs in Molecular Science, DOI: 10.1007/978-3-662-43624-0_1, © The Author(s) 2014

Table 1.1 Summary of nanomaterials used for optical nanoprobes

Novel nanomaterials				Strategy	Applications	Ref.
Metal						
Monocomponent	Gold	Nanostructure	Spheres	Aggregated; leaching	Colorimetry	[27, 28]
			Rods	Aggregated; etching	Colorimetry; SERS	[29]
			Popcorns	Modification	SERS	[10]
			Cluster (<10 nm)	Surface reaction	Fluorescence	[11, 12]
			Hollow shells	Modification	SERS	[10]
	Silver	Nanostructure	Spheres	Aggregated	Colorimetry; SERS	[13, 14]
			Rods	Substrate	SERS	[17]
			Nanocluster	Surface reaction	Fluorescence	[16]
			Prisms	Morphology transition	Colorimetry	[15]
Multicomponent	Nonmetal/metal nanostructure		Silica coated Au NRs	Surface change	Colorimetry	[30, 31]
			QDs	Aggregated; surface electron annihilation	Fluorescence	[3]
			Rare earth compounds	Surface electron annihilation	Fluorescence	[4–6]
	Metal/metal nanostructure		Au@Ag spheres	Leaching; aggregated	Colorimetry	[18]
			Au@Ag rods	Etching	Colorimetry	[19]
Nonmetal						
Graphene oxide (GO)				Assembly of aptamer with fluorochrome	Fluorescence	[7]
Carbon nanotube				Modification with fluorochrome	Fluorescence	[20]
Carbon dots				Surface reaction	Fluorescence	[21–23]
Assembly structures						
GO + Au NP/Au NRs				Etching; assembly	Colorimetry; fluorescence	[24, 25]
GO + Ag NPs				Assembly	SERS	[26]

Table 1.2 Summary of typical recognition units

Recognition units	Examples	Ref.
Small molecular ligands	Mercaptopropionic acid; thymine derivative; glutathione; L-Cysteine; 15-crown-5 moieties; etc.	[32–36]
Biological macromolecules	Aptamer; enzyme; antigen-antibody; etc.	[37–39]
Specific chemical reactions and interactions	Strong affinity between Au and Hg; click chemistry; Griess reaction; electrostatic interaction; Hydrogen-bonding; etc.	[40–44]

colorimetric and surface enhanced Raman scatting (SERS) methods [8–10], while smaller sized clusters were used for fluorescence response [11, 12]. Similarly, silver nanostructures played an important role in the construction of optical nanoprobes with varied optical signals [13–17]. Multicomponent nanostructures, such as QDs, rare earth compounds, offer better fluorescence properties in nano-sensors [3, 4, 6]; silver and gold hybrid spheres and rods support colorimetric methods [18, 19]. Graphene and carbon nanotube are usually used as substrate of probes with the aptamer-fluorochrome, because of their fluorescence quenching of fluorochrome [7, 20]. As a fluorescence material, carbon dot is often used in the biosystem for fluorescence imaging [21–23]. The assembly of two or more nanomaterials also could generate special nanoprobes for targets, such as the assembly of gold nanospheres or nanorods on the graphene surface [24–26]. With their unique optical properties, these novel nanomaterials can not only translate sensing behavior of target into optical signals but also endow high sensitivity of the nanosensor.

1.2 Recognition Units

The other key component in a nanoprobe is the recognition unit, which provides a selective response to the target. This unit generally can attach the surface of nanomaterial and simultaneously recognize target through specific interaction, skillfully combine the target sensing and the optical behavior of nanomaterial. The recognition units involve small organic molecules, biological macromolecule, and specific chemical/biological reaction and interaction. The recognition of target with these units induces the change of optical signals, such as color change, fluorescence, and SERS signals change.

Table 1.2 shows the demonstrated recognition units in recent reported optical nanoprobes. Most small molecular ligands have two key groups, one is thiol group, which is used for the linkage of nanosurface and N/O-group for the combination with target, and the other group is mainly used for the detection of metal ions, for instance, mercapto-aliphatic acid for heavy metal ions [32], thymine derivative for mercury ions [33, 34], glutathione for lead ions [35], 15-crown-5 moieties for potassium ions [36]. Biological macromolecules including aptamer, enzyme, and antigen-antibody can be used as outstanding recognition units in the optical nanosensors for the detection of special biomolecule and protein. The aptamer selected using the exponential enrichment (SELEX) methodology is an excellent recognition unit of nanosensor owning high selectivity toward target [37]. Based on the specific recognition between aptamer and target, aptamer-based nanoprobe can perform highly selective sensing behavior. Enzyme-based biological reactions also endow the nanoprobe high selectivity [38], as well as antigen–antibody [39]. Specific chemical reaction and interaction also can be used in the design of nanoprobes, including strong affinity between Au and Hg for the sensing of mercury ions [40], click chemistry for copper ions [41], Griess reaction for nitrite ions [42], electrostatic interaction [43], and hydrogen-bonding [44].

These recognition units determine the selectivity of the nanoprobes for the sensing of target. With their assistance, target-sensing behavior could be translated into optical signals.

1.3 Optical Detection

As the key components, nanomaterial and recognition units are essentially necessary in an optical assay, affecting the performance in selectivity, sensitivity, response time, and signal-to-noise ratio. In the sensing system, after the recognition behavior of target, the optical signals could be further detected by optical techniques.

The most commonly used optical detection techniques in nanoprobes include colorimetry, UV–Vis spectra, fluorescence spectra, and SERS. Generally, gold/silver nanomaterial-based nanoprobes could produce eye-sensitive color change in the visible range of 390–750 nm for colorimetric assay [45]. Fluorescence nanomaterial or fluorechrome mediated nanomaterial-based sensors usually employ fluorescence signals [4, 7]. Aggregated nanoparticles or rough nanosurface based sensor could produce "hot spot" for SERS tags with the increase of SERS signals [10]. These optical detection techniques realize the quantitative sensing of target.

In conclusion, an optical nanoprobe includes three key components, as shown in Fig. 1.1. As the nanosubstrates, novel nanomaterials with unique optical properties play the role of transducer moiety that can translate the sensing behavior into optical signals, which closely affect the sensitivity of nanosensor. The recognition units related to a range of specific ligands/reactions provide a selective response to the target. After the recognition behavior toward target, various optical detection techniques (colorimetry, UV–Vis spectra, fluorescence, SERS, etc.) are used according to the nanomaterial optical properties, to realize the detection of optical signals translated from target response. Thus, an optical nanoprobe becomes available for the detection of specific target in chemical and biological analysis.

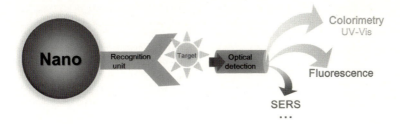

Fig. 1.1 The illustration of optical nanoprobe

References

1. Boisselier E, Astruc D (2009) Gold nanoparticles in nanomedicine: preparations, imaging, diagnostics, therapies and toxicity. Chem Soc Rev 38:1759–1782
2. Jain PK, Huang X, El-Sayed IH, El-Sayed MA (2008) Noble metals on the nanoscale: optical and photothermal properties and some applications in imaging, sensing, biology, and medicine. Acc Chem Res 41:1578–1586
3. Alivisatos AP, Gu W, Larabell C (2005) Quantum dots as cellular probes. Annu Rev Biomed Eng 7:55–76
4. Chatterjee DK, Gnanasammandhan MK, Zhang Y (2010) Small upconverting fluorescent nanoparticles for biomedical applications. Small 6:2781–2795
5. Haase M, Schäfer H (2011) Upconverting nanoparticles. Angew Chem Int Ed 50:5808–5829
6. Wang F, Banerjee D, Liu Y, Chen X, Liu X (2010) Upconversion nanoparticles in biological labeling, imaging, and therapy. Analyst 135:1839–1854
7. Liu Y, Dong X, Chen P (2012) Biological and chemical sensors based on graphene materials. Chem Soc Rev 41:2283–2307
8. Daniel M-C, Astruc D (2004) Gold nanoparticles: assembly, supramolecular chemistry, quantum-size-related properties, and applications toward biology, catalysis, and nanotechnology. Chem Rev 104:293–346
9. Rosi NL, Mirkin CA (2005) Nanostructures in biodiagnostics. Chem Rev 105:1547–1562
10. Wang Y, Yan B, Chen L (2013) SERS tags: novel optical nanoprobes for bioanalysis. Chem Rev 113:1391–1428
11. Lin Y-H, Tseng W-L (2010) Ultrasensitive sensing of Hg^{2+} and CH_3Hg^+ based on the fluorescence quenching of lysozyme type VI-stabilized gold nanoclusters. Anal Chem 82:9194–9200
12. Zhang J, Chen C, Xu X, Wang X, Yang X (2013) Use of fluorescent gold nanoclusters for the construction of a NAND logic gate for nitrite. Chem Commun 49:2691–2693
13. Ma YR, Niu HY, Zhang XL, Cai YQ (2011) Colorimetric detection of copper ions in tap water during the synthesis of silver/dopamine nanoparticles. Chem Commun 47:12643–12645
14. Rycenga M, Cobley CM, Zeng J, Li W, Moran CH, Zhang Q, Qin D, Xia Y (2011) Controlling the synthesis and assembly of silver nanostructures for plasmonic applications. Chem Rev 111:3669–3712
15. Xia Y, Ye J, Tan K, Wang J, Yang G (2013) Colorimetric visualization of glucose at the submicromole level in serum by a homogenous silver nanoprism-glucose oxidase system. Anal Chem 85:6241–6247
16. Yeh HC, Sharma J, Shih Ie M, Vu DM, Martinez JS, Werner JH (2012) A fluorescence light-up Ag nanocluster probe that discriminates single-nucleotide variants by emission color. J Am Chem Soc 134:11550–11558

17. Orendorff CJ, Gearheart L, Jana NR, Murphy CJ (2006) Aspect ratio dependence on surface enhanced Raman scattering using silver and gold nanorod substrates. Phys Chem Chem Phys 8:165–170

18. Lou T, Chen L, Chen Z, Wang Y, Chen L, Li J (2011) Colorimetric detection of trace copper ions based on catalytic leaching of silver-coated gold nanoparticles. ACS Appl Mater Interfaces 3:4215–4220

19. Wang X, Chen L, Chen L (2013) Colorimetric determination of copper ions based on the catalytic leaching of silver from the shell of silver-coated gold nanorods. Microchim Acta 181:105–110

20. Sinha N, Ma J, Yeow JT (2006) Carbon nanotube-based sensors. J Nanosci Nanotechnol 6:573–590

21. Cao L, Wang X, Meziani MJ, Lu F, Wang H, Luo PG, Lin Y, Harruff BA, Veca LM, Murray D (2007) Carbon dots for multiphoton bioimaging. J Am Chem Soc 129:11318–11319

22. Sun Y-P, Zhou B, Lin Y, Wang W, Fernando KS, Pathak P, Meziani MJ, Harruff BA, Wang X, Wang H (2006) Quantum-sized carbon dots for bright and colorful photoluminescence. J Am Chem Soc 128:7756–7757

23. Yang S-T, Cao L, Luo PG, Lu F, Wang X, Wang H, Meziani MJ, Liu Y, Qi G, Sun Y-P (2009) Carbon dots for optical imaging in vivo. J Am Chem Soc 131:11308–11309

24. Fu X, Chen L, Li J (2012) Ultrasensitive colorimetric detection of heparin based on self-assembly of gold nanoparticles on graphene oxide. Analyst 137:3653–3658

25. Fu X, Chen L, Li J, Lin M, You H, Wang W (2012) Label-free colorimetric sensor for ultrasensitive detection of heparin based on color quenching of gold nanorods by graphene oxide. Biosens Bioelectron 34:227–231

26. Ren W, Fang Y, Wang E (2011) A binary functional substrate for enrichment and ultrasensitive SERS spectroscopic detection of folic acid using graphene oxide/Ag nanoparticle hybrids. ACS Nano 5:6425–6433

27. Wang Z, Ma L (2009) Gold nanoparticle probes. Coordin Chem Rev 253:1607–1618

28. Chen Y-Y, Chang H-T, Shiang Y-C, Hung Y-L, Chiang C-K, Huang C-C (2009) Colorimetric assay for lead ions based on the leaching of gold nanoparticles. Anal Chem 81:9433–9439

29. Perezjuste J, Pastorizasantos I, Lizmarzan L, Mulvaney P (2005) Gold nanorods: synthesis, characterization and applications. Coordin Chem Rev 249:1870–1901

30. Wang G, Chen Z, Chen L (2011) Mesoporous silica-coated gold nanorods: towards sensitive colorimetric sensing of ascorbic acid via target-induced silver overcoating. Nanoscale 3:1756–1759

31. Wang G, Chen Z, Wang W, Yan B, Chen L (2011) Chemical redox-regulated mesoporous silica-coated gold nanorods for colorimetric probing of Hg^{2+} and S^{2-}. Analyst 136:174–178

32. Kim Y, Johnson RC, Hupp JT (2001) Gold nanoparticle-based sensing of "spectroscopically silent" heavy metal ions. Nano Lett 1:165–167

33. Chen L, Lou T, Yu C, Kang Q, Chen L (2011) N-1-(2-mercaptoethyl)thymine modification of gold nanoparticles: a highly selective and sensitive colorimetric chemosensor for Hg^{2+}. Analyst 136:4770–4773

34. Lou T, Chen L, Zhang C, Kang Q, You H, Shen D, Chen L (2012) A simple and sensitive colorimetric method for detection of mercury ions based on anti-aggregation of gold nanoparticles. Anal Method 4:488

35. Chai F, Wang C, Wang T, Li L, Su Z (2010) Colorimetric detection of Pb^{2+} using glutathione functionalized gold nanoparticles. ACS Appl Mater Interfaces 2:1466–1470

36. Lin S-Y, Liu S-W, Lin C-M, Chen C-h (2002) Recognition of potassium ion in water by 15-crown-5 functionalized gold nanoparticles. Anal Chem 74:330–335

37. Wang G, Wang Y, Chen L, Choo J (2010) Nanomaterial-assisted aptamers for optical sensing. Biosens Bioelectron 25:1859–1868

38. Wu Z, Wu ZK, Tang H, Tang LJ, Jiang JH (2013) Activity-based DNA-gold nanoparticle probe as colorimetric biosensor for DNA methyltransferase/glycosylase assay. Anal Chem 85:4376–4383

39. Zhen Z, Tang LJ, Long H, Jiang JH (2012) Enzymatic immuno-assembly of gold nanoparticles for visualized activity screening of histone-modifying enzymes. Anal Chem 84:3614–3620
40. Rex M, Hernandez FE, Campiglia AD (2006) Pushing the limits of mercury sensors with gold nanorods. Anal Chem 78:445–451
41. Zhou Y, Wang S, Zhang K, Jiang X (2008) Visual detection of copper(II) by azide- and alkyne-functionalized gold nanoparticles using click chemistry. Angew Chem Int Ed 47:7454–7456
42. Daniel WL, Han MS, Lee J-S, Mirkin CA (2009) Colorimetric nitrite and nitrate detection with gold nanoparticle probes and kinetic end points. J Am Chem Soc 131:6362–6363
43. Cao R, Li B, Zhang Y, Zhang Z (2011) Naked-eye sensitive detection of nuclease activity using positively-charged gold nanoparticles as colorimetric probes. Chem Commun 47:12301–12303
44. Ai K, Liu Y, Lu L (2009) Hydrogen-bonding recognition-induced color change of gold nanoparticles for visual detection of melamine in raw milk and infant formula. J Am Chem Soc 131:9496–9497
45. Du J, Jiang L, Shao Q, Liu X, Marks RS, Ma J, Chen X (2013) Colorimetric detection of mercury ions based on plasmonic nanoparticles. Small 9:1467–1481

Chapter 2
Colorimetric Nanoprobes

Abstract The development of highly sensitive, cost-effective, miniature nano-particle-based colorimetric nanoprobes attracted great attention in recent years. Depending on their excellent performance in environmental and biological analysis, colorimetric nanoprobes have been widely used for sensing a wide range of analytes/targets, such as metallic cations, anions, small organic molecules, oligonucleotides, proteins, cancer cells, etc. In this chapter, we first introduce the optical absorption properties of nanomaterial, mainly focusing on the noble metal nanomaterials, such as sphere gold nanoparticles, gold nanorods, and silver nanoparticles. Then we discuss the colorimetric sensing strategies for ions, small molecules, oligonucleotides, and protein detection and cellular analysis, highlighting some of their technical challenges and the new trends by means of a set of selected recent applications.

Keywords Noble metal nanomaterials · Optical nanoprobes · Gold nanoparticles · Silver nanoparticles · Colorimetric sensing

2.1 Optical Absorption Properties of Nanomaterials

As materials are reduced in size from the bulk to the nanoscale, they begin to exhibit new and unusual chemical and physical properties [1, 2]. Noble metallic nanostructures have attracted enormous scientific interest because of their unique size or shape dependent properties, including large optical field enhancements resulting in the strong scattering and absorption of light [3, 4]. Low-dimensional structures such as nanoparticles and nanostructured materials have excellent properties, such as quantum confinement of electrons and holes, surface effects, and geometrical confinement of phonons [4, 5].

In the last two decades, the interest in gold and silver metallic particles has dramatically increased, mostly because of their unique optical and electronic properties [3–8]. These unique properties are mainly due to the collective

L. Chen et al., *Novel Optical Nanoprobes for Chemical and Biological Analysis,*
SpringerBriefs in Molecular Science, DOI: 10.1007/978-3-662-43624-0_2,
© The Author(s) 2014

Fig. 2.1 Schematic diagrams illustrating localized surface plasmon resonance. Adapted with permission from Ref. [9]. Copyright 2014, Royal Society of Chemistry

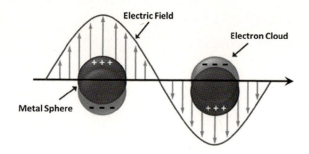

excitation of the conduction band electrons known as the surface plasmon resonance (SPR). The surface plasmon is a specific type of plasma oscillation occurring at lower energies than bulk plasmon, which happens when light is coupled to the coherent oscillation of free electrons at the surface of a conductor. This type of oscillation at resonant conditions is called the SPR. When the SPR is localized to a volume with dimensions smaller than the wavelength of incident light, it is called localized surface plasmon resonance (LSPR) [9].

When the particle dimensions are too small to support a propagating wave, light will interact with metal particles smaller than the wavelength of incident light to generate a LSPR (Fig. 2.1). The confinement of a surface plasmon to a small volume results in an oscillating electromagnetic field that resides very close to the particle surface, extending only nanometers into the dielectric environment [10]. Therefore, LSPR can generate much higher local field enhancements (100–10,000 times the incident field) comparing to those of SPP (10–100 times the incident field) [11]. The LSPR frequency can be tuned by changing the material composition, size, shape, and dielectric environment [12], which for gold, silver, and copper lies in the visible region [13].

Small spherical particles have a single, sharp absorption band due to the excitation of what is called a dipole plasmon resonance, where the entire charge distribution of the particle oscillates at the frequency of the incident electric field as illustrated in Fig. 2.1. For gold nanoparticles (Au NPs), the resonance condition is satisfied at visible wavelengths, which attributing for its intense color. Au NPs in the 10 nm size range have a strong absorption maximum around 520 nm in water due to their LSPR [9]. As shown in Fig. 2.2, Au NPs of different sizes present various color with different characteristic absorption bands.

Anisotropic particles can arise various LSPR modes. Nanorods are the quintessential demonstration of how optical properties are dependent on the dimensions (or shape) of a particle. The dipole plasmon resonance of a solution of nanorods is typically split between transverse and longitudinal dipole resonances due to the different dimensions along the width and length of the particles (Fig. 2.3A) [15]. The two bands positions depend on both the aspect ratio and the absolute dimensions of the particles [15]. The resonance at the longer wavelength (the longitudinal plasmon resonance) is associated with oscillations along the length of the nanorod, while the resonance at the shorter wavelength (the

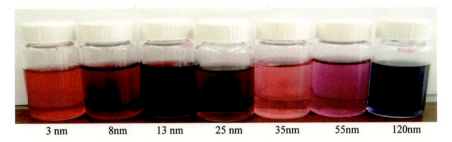

Fig. 2.2 Photograph showing gold nanoparticles of different sizes. Reprinted with permission from Ref. [14]. Copyright 2009, Wiley-VCH

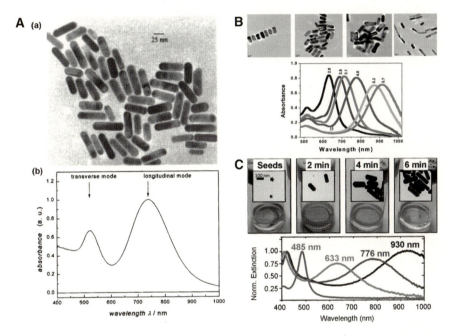

Fig. 2.3 **A** (a) TEM image of Au nanorods and (b) corresponding extinction spectrum. Due to the anisotropy of the particles, the dipole plasmon resonance is split into a transverse plasmon absorption at 525 nm and a longitudinal plasmon absorption at 740 nm. Adapted with permission from Ref. [15]. Copyright 1999, American Chemical Society. **B** TEM image of gold nanorods of average aspect ratios (σ) $\approx$ 2.0, 2.8, 4.0, and 5.2; Extinction profile of Au nanorods with aspect ratios varying from 2.0 to 5.7. The strong long wavelength band in the near-infrared region (λLPR = 600–950 nm) is due to the longitudinal oscillation of the conduction band electrons. The short wavelength peak ($\lambda \approx$ 520 nm) is from the nanorods' transverse plasmon mode. Reprinted with permission from Refs. [2, 18]. Copyright 2008, Wiley VCH. **C** Representative silver nanorod samples are shown as photographs and TEM images together with the corresponding ensemble extinction spectra (*bottom*). The samples correspond to Ag-seeds and Ag-nanorods grown with 2, 4, and 6 min of heating time (from *left* to *right*). Their plasmon peak increases from 485 to 633, 776, and 930 nm. Reprinted with permission from Ref. [17]. Copyright 2011, American Chemical Society

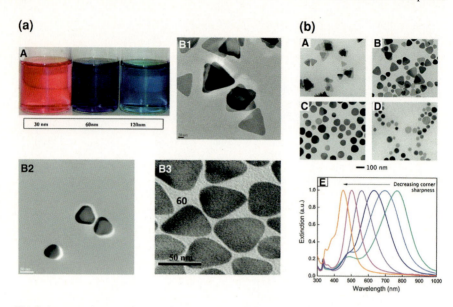

Fig. 2.4 a (A) Photographic images. (B1–B3) TEM pictures of silver nanoprisms of different sizes (30, 60, and 120 nm edge length). Reprinted with permission from Ref. [21]. Copyright 2009, Elsevier. **b** (A–D) TEM images of Ag *triangular nanoplates* with decreasing corner sharpness. The sharp cornered *triangular plates* in (A) were rounded until the particles had a *circular disk shape* (D). (E) The decreasing corner sharpness is correlated with a blue-shift in the extinction spectra of the nanoparticles in solution. Reproduced with permission from Ref. [20]. Copyright 2010, Wiley

transverse plasmon resonance) is associated with oscillations along the width of the nanorods [15, 16]. The LSPR of nanorods depend strongly on the aspect ratio. For example, Au NRs of different aspect ratio show various positions of two bands (Fig. 2.3B), the longer wavelengths exhibit red shift with the increasing of aspect ratio. The silver nanorods present the similar optical properties (seen in Fig. 2.3C). The absorption bands changed with the increasing of the aspect ratio [17].

The certain anisotropic particles can have extinction spectra that are greatly influenced by higher ordered resonances as evidenced by the in-plane and out-of-plane quadrupolar resonances observed for samples of silver triangular nanoprisms [19, 20]. As shown in Fig. 2.4a, with the increasing of edge length, silver nanoparisms showed red to blue color change. The decreasing of corner sharpness is correlated with a blue shift in the extinction spectra of silver nanoprisms (Fig. 2.4b).

The various compositions of nanoparticles could cause the change of SPR frequency with visible changes to the color of a colloid [22, 23]. As shown in Fig. 2.5A, the addition of sulfide ions into the Au@Ag core–shell nanocubes generated stable Au@Ag/Ag_2S core–shell nanoparticles at room temperature, and the plasmon extinction maximum shifts to the longer wavelength covering the entire visible range of 500–750 nm with full-color tuning [24]. The silver-coated

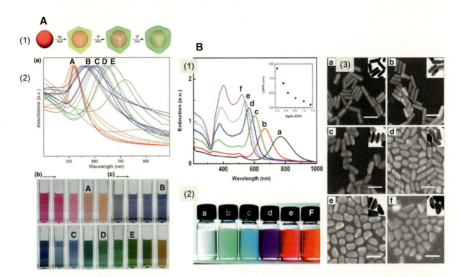

Fig. 2.5 **A** (1) Synthesis of Au@Ag/Ag₂S core–shell nanoparticles surface plasmon resonance shifts of Au seeds, Au@Ag core–shell nanocubes, and Au@Ag/Ag₂S nanoparticles. (2) (a) UV-vis absorption spectra. Photographs of the aqueous dispersions of (b) Au seeds and Au@Ag core–shell nanocubes, and (c) Au@Ag/Ag₂S nanoparticles. The representative samples are referred to as A (*orange*), B (*dark blue*), C (*blue*), D (*bluish green*), and E (*green*). Reprinted with permission from Ref. [24]. Copyright 2012, American Chemical Society. **B** (1) UV-vis extinction spectra of (a) Au NRs and (b–f) the Au@Ag core/shell nanocrystals with Ag/Au molar ratios. The *inset* in panel A shows the relationship between the longitudinal SPR position and the Ag/Au molar ratio. (2) Photographs of nanocrystal dispersions corresponding to the *curves* in panel 1. (3) SEM images of (a) Au NRs and (b–f) Au@Ag core/shell nanocrystals with Ag/Au molar ratios of 0.28, 0.49, 0.83, 1.20, and 1.51 measured by EDX. The *scale bar* is 60 nm. *Insets* are corresponding STEM images with the same *scale bar*. Reproduced with permission from Ref. [23]. Copyright 2012, American Chemical Society

gold nanorods from homogeneous coating to anisotropic coating also exhibit peak position shift [23]. As shown in Fig. 2.5B, with increasing overall amount of the deposited Ag, the longitudinal SPR band blue shifts obviously, while the transverse SPR band blue shifts slightly and is finally smeared out by the former.

Furthermore, the SPR frequency is sensitive to the proximity of other nanoparticles. When two or more discrete plasmonic materials are in close proximity to one another (on the order of nanometers), their oscillating electric fields can interact to yield new resonances and the surface plasmons will be coupled [25]. For instance, the aggregation of Au NPs results in significant red-shifting (from ∼520 to ∼650 nm for 13 nm Au NPs) and broadening in the surface plasmon band, changing the solution color from red to blue due to the interparticle plasmon coupling [3]. This visibly apparent phenomenon has made Au NPs an attractive candidate for colorimetric sensors [7–26].

Colorimetric assay is that the molecular recognition event can be transformed into color change, which can be easily observed by the naked eye. Depending on their size, shape, degree of aggregation, and sensitive surface, novel metallic nanoparticles can appear abundant colors and emit bright resonance light scattering of various wavelengths. Therefore, spherical nanoparticles and nonspherical structures (e.g., prisms, rods, cubes) with unique shape-dependent properties can be extensively explored as colorimetric probes with the help of recognition moiety for sensing a wide range of analytes/targets, such as metallic cations, nucleic acids, proteins, cells, etc. [5, 7]. Owing to the excellent optical properties, gold and silver nanostructures are most attractive for optical applications.

2.2 Colorimetric Sensing Strategies

Nanomaterial-based colorimetric methods are commonly based on the change of optical properties due to assemblies (or aggregations), morphology transition, and the surface chemical reaction, accompany by distinct color change. Two key components are essentially necessary in a colorimetric assay and affect the performance in selectivity, sensitivity, response time, and signal-to-noise ratio. One is the recognition moiety that provides a selective/specific response to the analyte, which relates to a wide range of organic or biological ligands/reactions. The other is the transducer moiety that translates detecting behavior into an eye-sensitive color change in the visible range of 390–750 nm [27]. Based on the behavior of nanoparticles, colorimetric sensing strategies can be summarized as two types: one is based on the analyte-induced aggregation of nanoparticles, called aggregation-based method. The other mainly uses the morphology transition and the surface chemical reaction of single particle, which is called single particle morphology-based method.

2.2.1 Aggregation-Based Methods

The aggregation of noble metal nanoparticles (Au/Ag) of appropriate sizes induces interparticle surface plasmon coupling, resulting in a visible color change. For instance, the aggregation of Au NPs (d > 3.5 nm) induces interparticle surface plasmon coupling, resulting in a visible color change from red to blue at nanomolar concentrations [7, 28, 29]. The color change during nanoparticle aggregation provides a practical platform for absorption-based colorimetric sensing of any target analyte that directly or indirectly triggers the nanoparticles aggregation.

Aggregation-based colorimetric nanoprobes have been widely explored [5]. Generally, nanoparticles are modified different types of functional decorators such as oligonucleotides, aptamer, functional molecules (Fig. 2.6), the presence of target analyte can induces the aggregation by specific interaction with functional ligands, resulting in two types of aggregation: crosslinking aggregation and non-crosslinking aggregation.

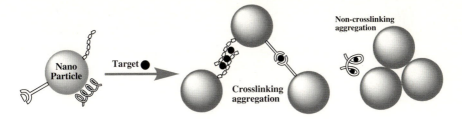

Fig. 2.6 Schematic illustration of aggregation-based colorimetric nanoprobe

Crosslinking aggregation-based colorimetric sensing methods generally require the incorporation of chelating agents onto the nanoparticle surface. The presence of target analyte induces the nanoparticle aggregation by forming multidentate inter-particle complexes with the chelating ligand (Fig. 2.6). Thiol is one of the main functional groups for tethering to the surface of nanoparticles due to the strong Au–S or Ag–S bond. Therefore, mercapto derivatives of ligands become preferred deco-rators for a nanoprobe system. The incentives of crosslinking aggregation in nano-particle-based colorimetric methods involve chelate reaction, chemical interaction, electrostatic interaction, base paring, hydrogen-bonding recognition, and so on.

For instance, Chen et al. developed an Au NP-based colorimetric assay for mercury ions (Hg^{2+}) detection using the coordination reaction between thymine (T) and Hg^{2+} [30]. The synthesized N-1-(2-mercaptoethyl) thymine can be easily coupled to the surface of Au NPs through Au–S bond. The presence of Hg^{2+} could effectively induce the aggregation of Au NPs by forming a T–Hg–T complex with strong affinity owing to the active N site, resulting in a significant color change from red to blue (Fig. 2.7a). This is a typical crosslinking aggregation-based strategy using the chelate reaction between ligands and metal ions. Jiang's group reported a method for the detection of copper ions (Cu^{2+}) by azide-and terminal alkyne-functionalized Au NPs in aqueous solutions using click chemistry [31]. When Cu^{2+} and the reductant (sodium ascorbate) were both introduced to the mixture of azide- and alkyne- functionalized Au NPs, Cu (I) could catalyze 1, 3-dipolar cyclo-addition of alkynes and azides on the surface of functionalized Au NPs, resulting in the aggregation of Au NPs (Fig. 2.7b). Cao et al. explored a simple and visual approach to colorimetric detection of nuclease activity using the electrostatic interaction between positively charged Au NPs and negatively charged DNA [32]. The polyanionic ssDNA used as the S1 nuclease substrate could induce the aggregation of (+) Au NPs, which can be observed via the color change from red to blue. During the enzymatic reaction of S1 nuclease, ssDNA is degraded to smaller fragments, which cannot induce the color change (Fig. 2.7c). Thus, electrostatic interaction-induced aggregation of Au NP-based colorimetric method was achieved. DNA-base paring meditated Au NPs assembly was dem-onstrated by Mirkin's group [33]. In this method, two thiolated ssDNA modified Au NP probes were fabricated for colorimetric detection of target oligonucleotides (Fig. 2.7d). Upon the addition of target, Au NPs aggregation was trigged with

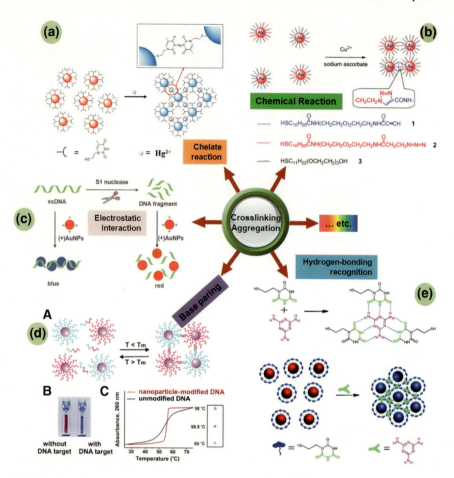

Fig. 2.7 Illustration of the crosslinking aggregation-based colorimetric nanoprobes. **a** Schematic mechanism of T–S–Au NPs sensing Hg^{2+} in aqueous solutions based on the Hg^{2+}-induced aggregation of gold nanoparticles. Reprinted with permission from Ref. [30]. Copyright 2011, Royal Society of Chemistry. **b** The detection of Cu^{2+} ions using click chemistry between two types of gold NPs, each modified with thiols terminated in an alkyne (1) or an azide (2) functional group. Reprinted with permission from Ref. [31]. Copyright 2008, Wiley. **c** Colorimetric assay of nuclease activity based on the color change of the positively charged gold nanoparticles ((+)AuNPs). Charge interaction between (+) AuNPs and polyanionic ssDNA leads to the aggregation of the AuNPs, which can be observed via the color change from *red* to *blue*. During the enzymatic reaction of S1 nuclease, ssDNA is degraded to smaller fragments, and the smaller fragments cannot induce the color change. Reprinted with permission from Ref. [32]. Copyright 2011, Royal Society of Chemistry. **d** In the presence of complementary target DNA, oligonucleotide-functionalized gold nanoparticles will aggregate (A), resulting in a change of solution color from *red* to *blue* (B). The aggregation process can be monitored using UV-vis spectroscopy or simply by spotting the solution on a silica support (C). Reprinted with permission from Ref. [35]. Copyright 2005, American Chemical Society. **e** Hydrogen-bonding recognition between melamine and cyanuric acid derivative. Colorimetric detection of melamine using the MTT-stabilized gold nanoparticles. Reprinted with permission from Ref. [34]. Copyright 2008, American Chemical Society

Fig. 2.8 **a** Schematic representation of Hg^{2+} stimulated aggregation of AuNPs and **b** TEM images of non-aggregated AuNPs stabilized by ssDNA in the presence $NaClO_4$ and the aggregated Au NPs after addition of Hg^{2+}. Reproduced with permission from Ref. [36]. Copyright 2008, Wiley

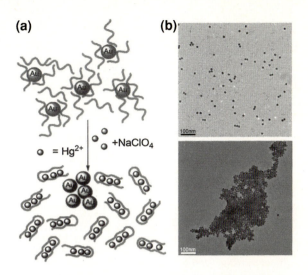

concomitant color change as a result of hybridization of the DNA strand. Highly specific base-pairing of DNA strands coupled with the intense absorptivity of Au NPs enables the subpicomolar quantitative colorimetric detection of oligonucleotides. Lu's group developed a hydrogen-bonding recognition-induced color change of Au NPs for visual detection of melamine in raw milk and infant formula [34]. Upon exposure of the synthesized 1-(2-mercaptoethyl)-1, 3, 5-triazinane-2, 4, 6-trione (MTT)-stabilized Au NPs to melamine, hydrogen-bonding recognition between melamine and MTT resulted in the aggregation of Au NPs, and the wine red color was accordingly changed to a blue color (Fig. 2.7e).

Another type of aggregation-based method relies on noncrosslinking aggregation of nanoparticles in different modes. Generally, free specific ligands are treated with nanoparticles through weaker interaction. Macroscopic changes of nanoparticles were attributed to the stronger interaction between ligands and target analyte, and following release of ligands from the surface of nanoparticles. The system became so destabilized that nanoparticles aggregated. For instance, Willner and coworkers fabricated a simple, fast, and wide-range system for the detection of Hg^{2+} using T rich-nucleic acid [36]. After treating Au NPs (13 nm) with ssDNA, $NaClO_4$ (100 mM) was added into the solution to maintain a high level of salinity. Under this condition, red color with an absorption band at 520 nm was observed before addition of Hg^{2+}, indicating good dispersion of the nanoparticles. Upon addition of Hg^{2+} the solution turned blue, induced by aggregation of Au NPs, which was detected by the naked eye and also confirmed by a red shift and broadened peak in the UV-vis spectrum. Macroscopic changes were attributed to the formation of the nucleic acid duplex-folded complex with the help of Hg^{2+}, and following release of ssDNA from Au NPs. As a result, the system became so destabilized that Au NPs aggregated (Fig. 2.8).

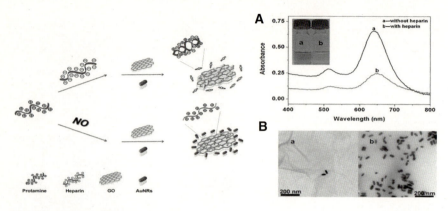

Fig. 2.9 Schematic illustration of GO quenching the color of Au NRs sensing heparin based on the self-assembly of CTAB-stabilized Au NRs on the surface of GO. **A** Absorption spectra and **B** TEM images of Au NRs in the GO/protamine mixed solutions in the absence (a) and presence (b) of heparin (0.24 μg/mL). *Inset* of part **A**: Photographic images of the corresponding colorimetric response (*scale bars* 200 nm). Reprinted with permission from Ref. [37]. Copyright 2011, Elsevier

A novel label-free colorimetric strategy was developed for ultrasensitive detection of heparin by using the super color quenching capacity of graphene oxide (GO) by Fu et al. [37]. Hexadecyltrimethylammonium bromide (CTAB)-stabilized Au NRs could easily self-assemble onto the surface of GO through electrostatic interaction, resulting in the decrease of the SPR absorption and consequent color quenching change of Au NRs from deep to light. Polycationic protamine was used as a medium for disturbing the electrostatic interaction between Au NRs and GO, as shown in Fig. 2.9. The Au NRs were prevented from adsorbing onto the surface of GO because of the stronger interaction between protamine and GO, showing a native color of Au NRs. On the contrary, in the presence of heparin, which was more easily to combine with protamine, the Au NRs could self-assemble onto the surface of GO, resulting in the native color disappearing of Au NRs. The amounts of self-assembled Au NRs were proportional to the concentration of heparin, and thereby the changes in the SPR absorption and color had been used to monitor heparin levels. A good linearity was obtained in a range of 0.02–0.28 μg/mL, and a limit of detection (LOD) was 5 ng/mL.

Redox formed metal coating on the surface of nanoparticles is an excellent analytical strategy for noncrosslinking-aggregation based colorimetric assay. Lou et al. developed a "blue-to-red" colorimetric method for determination of mercury ions (Hg^{2+}) and silver ions (Ag^+) based on stabilization of Au NPs by redox formed metal coating in the presence of ascorbic acid (AA) [38]. In this method, Au NPs were first stabilized by Tween 20 in phosphate buffer solution with high ionic strength. In a target ion-free system, the addition of N-acetyl-L-cysteine resulted in the aggregation of Tween 20 stabilized Au NPs for mercapto-ligand self-assembled on the surface of Au NPs, which induced the Au NPs to be

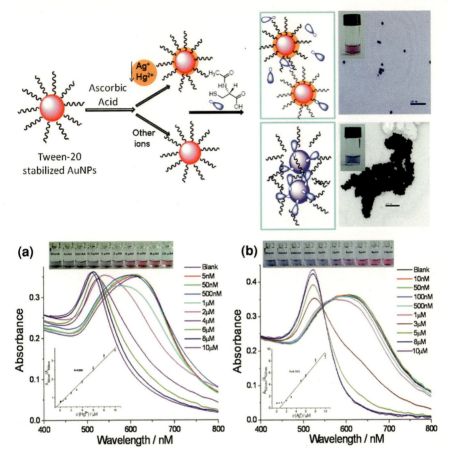

Fig. 2.10 Illustration of colorimetric sensing mechanism based on redox reaction modulated surface chemistry of Au NPs; TEM image of aggregated Au NPs in Hg^{2+}-free PBS solution and the monodispersed Au NPs Obtained after Addition of 10 μM Hg^{2+}. Digital photographs, absorption spectra, and plots of $A_{520\ nm}/A_{620\ nm}$ versus the concentration (*inset*) in 50 mM PBS (pH 7.2) containing Tween 20-Au NPs, 1 mM AA, and 0.1 M NaCl (0.01 M EDTA) upon addition of **A** 0–10 μM Hg^{2+} and **B** Ag^+. The incubation time was 3 min. The error bars represented standard deviations based on three independent measurements. Reprinted with permission from Ref. [38]. Copyright 2011, American Chemical Society

unstable. This would lead to a color change from red to blue. By contrast, in an aqueous solution with Hg^{2+} or Ag^+, the ions could be reduced with the aid of AA to form Hg–Au alloy or Ag coating on the surface of Au NPs. This metal coating blocked mercapto-ligand assembly and Au NPs kept monodispersed after addition of N-acetyl-L-cysteine, exhibiting a red color, as shown in Fig. 2.10. This method could selectively detect Hg^{2+} and Ag^+ as low as 5 and 10 nM with a linear range of 0.5–10 μM for Hg^{2+} and 1.0–8.0 μM for Ag^+, respectively.

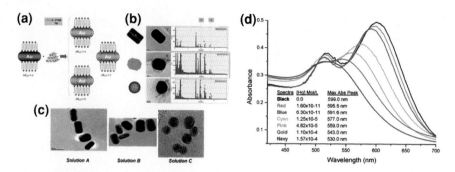

Fig. 2.11 Schematic diagram showing the amalgamation of Hg with Au nanorods. **B** TEM and EDX analysis of Au nanorods in the absence and the presence of Hg. I = no Hg; II = 1.25×10^{-5} M and III = 1.57×10^{-4} M Hg^{2+}. All solutions were prepared in 1.67×10^{-3} mol/L NaBH$_4$. **C** TEM analysis of multiple Au nanorods in the absence and the presence of Hg. Sol A, no Hg; Sol B, 1.25×10^{-5} M Hg; Sol C, 1.57×10^{-5} M Hg^{2+}. All solutions were prepared in 1.67×10^{-3} mol/L NaBH$_4$. UV-visible absorption spectra showing the spectral shift at several Hg (II) concentrations. The concentration range between 1.6×10^{-11} and 6.3×10^{-11} M shows the spectra within the linear dynamic range of the calibration curve. The remaining spectra show the overlapping between the longitudinal and transversal absorption bands at higher Hg (II) concentrations. Reproduced with permission from Ref. [41]. Copyright 2006, American Chemical Society

2.2.2 Single Particle Morphology-Based Methods

The color-tunable behavior of nanoparticles depends on the size and shape enables nanomaterials an attractive candidate for the colorimetric nanoprobes. Many kinds of nanomaterials involve gold nanoparticles (spheres, rods), silver-coated gold nanoparticles, silver nanoprisms have been used for colorimetric probes [39, 40], relying on the analyte-induced morphology transition, and the surface chemical reaction of single particle, calling the single particle morphology-based methods.

Campiglia et al. provided a direct way to determine mercury using the morphology change of Au NRs resulted from the well-known amalgamation process that occurs between mercury and gold [41]. With the addition of Hg^{2+} and reductant (NaBH$_4$) to the Au NRs solution, chemical reactions involved in the amalgamation of mercury and gold gradually occurred at the active sites of Au NRs-the tips of nanostructures, that could cause a reduction of effective aspect ratio of the nanorods and a blue shift of the maximum absorption wavelength of the longitudinal mode band (Fig. 2.11).

Lou et al. explored a colorimetric, label-free, and nonaggregation-based silver-coated gold nanoparticles (Ag/Au NPs) probe for detection of trace Cu^{2+} in aqueous solution, based on the fact that Cu^{2+} can accelerate the leaching rate of Ag/Au NPs by thiosulfate (S$_2$O$_3^{2-}$) [42]. The leaching of Ag/Au NPs would lead to dramatic decrease in the SPR absorption as the size of Ag/Au NPs decreased (Fig. 2.12). This colorimetric strategy based on size-dependence of nanoparticles during their leaching process provided a highly sensitive (1.0 nM) and selective

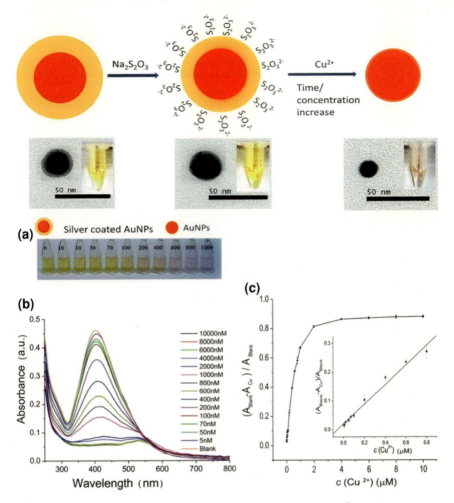

Fig. 2.12 Schematic representation of the sensing mechanism of the $S_2O_3^{2-}$–Ag/AuNPs for the colorimetric detection of Cu^{2+} based on catalytic leaching of Ag-coated Au NPs. **A** Photographs and **B** absorption responses of the $S_2O_3^{2-}$–Ag/AuNP solution addition of Cu^{2+}, and **C** plot of $(A_{blank}-A_{Cu})/A_{blank}$ (at 405 nm) values versus Cu^{2+} concentration. *Inset* the enlarged portion of the plot in the Cu^{2+} concentration range of 5–800 nM; the regression equation is $(A_{blank}-A_{Cu})/A_{blank} = 0.0202 + 0.341c$ (μM) ($r = 0.991$). Reprinted with permission from Ref. [42]. Copyright 2011, American Chemical Society

detection toward Cu^{2+}, with a wide linear detection range (5–800 nM) over nearly three orders of magnitude. Wang et al. also developed a silver-coated gold nanorod-based Cu^{2+} probe by the same reaction principle [43]. A sensitive colorimetric sensing method for AA was also developed via target-induced silver overcoating using mesoporous silica-coated Au NRs [44].

Yang's group designed a homogeneous system consisting of Ag nanoprisms and glucose oxidase (GOx) for simple, sensitive, and low-cost colorimetric sensing of

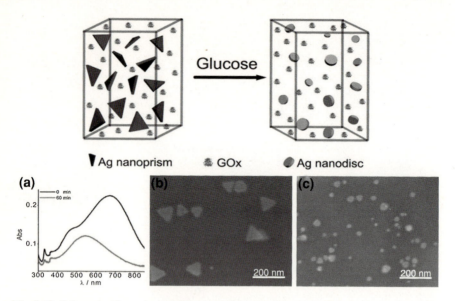

Fig. 2.13 Schematic illustration of the (Ag nanoprism)-GOx homogenous system for colorimetric sensing of Glucose. SPR absorption spectra (**A**) and SEM images of the Ag nanoprisms before (**B**) and after (**C**) incubation with GOx and glucose (100 μ M) for 60 min. Reprinted with permission from Ref. [45]. Copyright 2013, American Chemical Society

glucose in serum [45]. In this study, the unmodified Ag nanoprisms and GOx are first mixed with each other. Glucose is then added in the homogeneous mixture. Finally, the nanoplates are etched from triangle to round by H_2O_2 produced by the enzymatic oxidation (Fig. 2.13), which leads to a more than 120 nm blue shift of the SPR absorption band of the Ag nanoplates. This large wavelength shift can be used not only for visual detection (from blue to mauve) of glucose by naked eyes but also for reliable and convenient glucose quantification in the range from 2.0×10^{-7} to 1.0×10^{-4} M. The detection limit is as low as 2.0×10^{-7} M, because the used Ag nanoprisms possess not only highly reactive edges/tips but also strongly tip sharpness and aspect ratio dependent SPR absorption. Owing to ultrahigh sensitivity, only 10–20 μL of serum is enough for a one-time determination.

2.3 Applications

2.3.1 Ionic Detection

Colorimetric nanoprobes have been widely developed for ionic detection in environmental and biological analysis, including alkali and alkaline earth metal ions, heavy metal ions, anions. These colorimetric sensing methods were mainly fabricated using novel metal nanoparticles. Table 2.1 summarized the key characteristics of colorimetric nanoprobes for ionic detection.

Table 2.1 Key characteristic of colorimetric nanoprobes for target analysis

Target/Analyte	Nanomaterial	Detection linear range	Sensing strategy	References
Metal ions				
Hg^{2+}	Au NPs	0.1–2.0 μM	Hg^{2+} will coordinate selectively to the bases that make up a T–T mismatch	[46]
	Au NPs	250–500 nM	Hg^{2+} induce the aggregation of mercaptopropionic acid-modified Au NPs	[47]
	Au NRs	0.016 nM–157 μM	Hg^{2+} was reduced by sodium borohydride to form amalgamation and cause a reduction of the aspect ratio of AuNRs	[41]
	Au NPs	0.5–50 μM	Hg^{2+} induces the aggregation of DNA/AuNPs	[48]
	Au NPs	10–300 nM	Hg^{2+} induces the non-crosslinking aggregation of T_{33}-DNA/Au NPs	[49]
	Au NPs	5–1,000 nM	Hg^{2+} bridged crosslinking aggregation of N-1-(2-mercaptoethy) thymine modified Au NPs	[30]
	Au NPs	0.2–6.0 μM	Hg^{2+} induces the non-crosslinking aggregation of mononucleotides-stabilized Au NPs	[50]
	Au NPs	2–12 μM	Hg^{2+} inhibits the thymine-induced aggregation of Au NPs	[51]
	Ag NPs	2–1,000 nM	Catalytic reduction property of Ag NPs	[52]
	Silver nanoprisms	10–500 nM	Hg^{2+} indirectly induce the morphology transition of silver nanoprisms	[53]
Hg^{2+}, Ag^+	Au NPs	200–800 nM for Hg^{2+} 400–1,000 nM for Ag^+	Hg^{2+} and Ag^+ were reduced by citrate to form Hg–Au alloys and Ag on the surface of the Au NPs	[54]
	Au NPs	0.5–10 μM for Hg^{2+} 1.0–8.0 μM for Ag^+	Blue-to-red colorimetric sensing for Hg^{2+} and Ag^+ via redox-regulated surface chemistry of Au NPs	[38]
Hg^{2+}, Pb^{2+}, Cu^{2+}	Au NPs	4–28 μM	The interaction between protein-functionalized Au NPs and metal ions	[55]

(continued)

Table 2.1 (continued)

Target/Analyte	Nanomaterial	Detection linear range	Sensing strategy	References
Pb^{2+}	Au NPs	2.5 Nm–10 μM	Pb^{2+} ions accelerate the leaching rate of Au NPs by thiosulfate ($S_2O_3^{2-}$) and 2-mercaptoethanol (2-ME)	[39]
	Au NPs	3 nM–1 μM (pH 7.2) 120 nM–20 μM (pH 5.5)	AuNPs aggregate in the absence of lead but remain dispersed in the presence of lead by the addition of DNAzyme and NaCl	[56]
	Au NPs (5–8 nm)	0.1–50 μM	Pb^{2+} induce the aggregation of glutathione-stabilized Au NPs	[57]
As^{3+}	Au NPs	0–450 ppt	As^{3+} induces the crosslinking aggregation of GSH/DTT/Cys-modified Au NPs	[58]
Cd^{2+}	Au NPs	0.2–1.7 μM	Cd^{2+}-induced the aggregation of 6-mercaptonicotinic acid and L-Cysteine co-functionalized Au NPs	[59]
Cr^{3+}	Au NPs	0.1–1.0 μM	N-benzyl-4-(pyridin-4-ylmethyl)aniline modified Au NPs aggregated in the presence of Cr^{3+}	[60]
Cr (VI)	Au NRs	0.1–20 μM	Selective etching of Au NRs at tips	[61]
Co^{2+}	Au NPs	0.1–0.7 μM	Co^{2+} induce aggregation of thiosulfate stabilized gold nanoparticles in the presence of ethylenediamine	[62]
Cu^{2+}	Au NPs	50–500 μM	Assay for Cu^{2+} by the aggregation of azide and alkyne functionalized Au NPs as a result of the Cu(I)-catalyzed conjugation	[31]
	Au@Ag NPs	5–800 nM	Cu^{2+} accelerate leaching of silver-coated gold nanoparticles	[42]
	Ag NPs	3.2–512 ppb	Cu^{2+} can interact with dopamine during the synthesis of silver/dopamine nanoparticles	[63]
	Au NRs	10–300 nM	Cu^{2+} accelerate decomposition of H_2O_2 using Au NRs	[64]
	Au@Ag NRs	3–1,000 nM	Cu^{2+} accelerate the leaching of silver-coated gold nanorods	[43]
	Au NRs		Cu^{2+} accelerate the etching of Au NRs by dissolved oxygen in the presence of NH_3–NH_4Cl	[65]

(continued)

Table 2.1 (continued)

Target/Analyte	Nanomaterial	Detection linear range	Sensing strategy	References
Fe^{3+}	Au NPs	10–60 μM	Fe^{3+} induce the aggregation of pyrophosphate functionalized Au NPs	[66]
K^+	Au NPs	7.6–140 μM	15-crown-5 moieties functionalized Au NPs	[67]
	Au NPs	6.25–112 μM	15-crown-5-$CH_2O(CH_2)_{12}SH$ functionalized Au NPs	[68]
Ca^{2+}	Au NPs	0.1–1.6 mM	Cytidine triphosphate stabilized gold nanoparticles	[69]
	Au NPs	1.9–20 μM	1-thiohexyl carboxylic acid and 1-thiohexyl β-D-lactopyranoside in bi-ligand functionalized AuNPs	[70]
Eu^{3+}	Au NPs	50–496 nM	Crosslinking aggregation of tetramethylmalonamide (TMMA) functionalized Au NPs	[71]
Anions				
NO_2^-	Au NPs	20–35 μM	Nitrite coupled Griess reaction between aniline Au NPs and naphthalene Au NPs	[72]
	Au NRs	5.2–60 μM	Nitrite trigger the non-crosslinking aggregation of 4-aminothiophenol modified Au NRs	[73]
	Au NRs	1.0–15 μM	Nitrite trigger the etching of Au NRs at the ends	[74]
HCl	Au NPs	500 ~ 5,000 ppm	Nonaggregation-based detection system relies on the ability of chloro species to cause rapid leaching of Au NPs in an aqueous dispersion containing a strong oxidizing agent	[75]
I^-	Au NPs	10–600 nM	Anti-aggregation of Au NPs	[76]
SCN^-	Au NPs	0.2–2 μM	SCN^- induce the non-crosslinking aggregation of Tween 20-stabilized Au NPs	[77]
Small organic molecules				
TNT	Au NPs	0.5 pM–5 nM	TNT-induced the aggregation of cysteamine stabilized Au NPs	[78]
	Au NPs	50–250 μM	TNT trigger the aggregation of p-ATP attached Au NPs	[79]

(continued)

Table 2.1 (continued)

Target/Analyte	Nanomaterial	Detection linear range	Sensing strategy	References
Glucose	Ag nanoprism	$0.2 \sim 100$ μM	Silver nanoprisms are etched from triangle to round by H_2O_2 produced by the enzymatic oxidation	[45]
Dopamine	Au NPs	10–90 μg/mL	Glucose oxidase immobilized Au NPs aggregate in the presence of glucose	[80]
	Au NPs	5–350 nM	Dopamine induces the aggregation of both 4-mercaptophenylboronic acid (MBA) and dithiobis(succinimidylpropionate) (DSP) functionalized the Au NPs	[81]
Melamine	Au NPs	0.2–1.1 μM	Hydrogen-bonding recognition induced aggregation of 4-amino-3-hydrazino-5-mercapto-1,2,4-triazol (AHMT) functionalized Au NPs	[82]
	Au NPs	0–400 nM	Hydrogen-bonding recognition-induced color change of 1-(2-mercaptoethyl)-1,3,5-triazinane-2,4,6-trione (MTT) modified Au NPs	[34]
	Au NPs	0.6–1.6 μM	Melamine can induced the aggregation of Au NPs and results in the color change from wine-red to purple	[83]
Ascorbic acid	Au NPs	$4.4 \sim 30$ nM	Alkyne–azide click reaction using gold nanoparticles with the addition of Cu^{2+}	[84]
	Mesoporous silica coated Au NRs	0.1–2.5 μM	Tailoring the optical properties of mesoporous silica-coated Au NRs via silver overcoating	[44]
ATP	Au NPs	$0.6 \sim 132.7$ μM	Au NP-based aptamer target binding readout for ATP assay	[85]
	Au NPs	0.1–1 mM	Adenosine induces assembly of aptazyme functionalized Au NPs	[86]
	Au NPs	0.3–2 mM	Aptamer attached Au NPs	[87]
Cocaine	Au NPs	50–500 μM	Cocaine binding aptamer attached Au NPs	[87]
	Au NPs	2–200 nM	Cocaine can interact with engineered aptamer and induce the aggregation of Au NPs	[88]
Streptomycin	Au NPs	2 nM ~ 1.8 μM	Streptomycin induces the aggregation of MPA-AuNPs based on the electrostatic interaction between streptomycin and MPA	[89]

(continued)

Table 2.1 (continued)

Target/Analyte	Nanomaterial	Detection linear range	Sensing strategy	References
Clenbuterol	Au NPs	0.28–280 nM	Hydrogen-bonding interaction between clenbuterol and melamine result in the aggregation of Au NPs	[90]
Oxytetracycline	Au NPs	280–1,400 nM, 1–200 μM	Oxytetracycline bound aptamer and Au NPs aggregate in the presence of salt	[91]
Organophosphate pesticide	Au NPs	2.51–18.7 pM for VX, 15.0–52.5 pM for GD, 28.2–225 pM for GB, 45.2–495 fM for paraoxon	Catalytic reaction of acetylcholine esterase (AChE) and the aggregation of lipoic acid capped AuNPs	[92]
Phthalates	Au NPs	1–10,000 ppm	UTP-modified gold nanoparticles cross-linked by copper(II)	[93]
Cysteine	Au NPs	0.01–100 μM	Self-assembly of cysteine on gold nanoparticles in the presence of copper ions	[94]
Cysteine and glutathione	Au NRs	1.75–3 μM for Cys, 7–14 μM for GSH	Cysteine and glutathione induce assembly of Au NRs	[95]
D-Cys	Ag NPs	0.1–20 μM	UTP-capped Ag NPs can be used as an ultrahigh efficiency enantioseparation and detection platform for D-and L-cysteine	[96]
Oligonucleotides				
Single nucleotide polymorphisms	Au NPs	2–80 nM	Single-stranded DNA binding protein–nucleic acids interaction and unmodified Au NPs	[97]
Nucleic acids	Au NPs	2–10 nM	Nonionic morpholino oligos functionalized Au NPs	[98]
Proteins				
Nuclease	Au NPs	0 ~ 30 U/mL	Charge interaction between positively-charged AuNPs and long DNA	[32]

(continued)

Table 2.1 (continued)

Target/Analyte	Nanomaterial	Detection linear range	Sensing strategy	References
Thrombin	Au NPs	2 ~ 167 nM	Aptamer-functionalized Au NPs for the amplified optical detection of thrombin	[99]
Histone-modifying enzymes	Au NPs	1–200 nM	Antibody-mediated assembly of Au NPs decorated with substrate peptides	[100]
Platelet-derived growth factors	Au NPs	2.5–200 nM	Aptamer modified Au NPs aggregate in the presence of target	[101]
Trypsin and inhibitor	Au NPs	1.6–8.0 ng/mL	Using crosslinking/aggregation of Au NPs based on trypsin-catalyzed hydrolysis of Arg6	[102]
Acetylcholinesterase	Au NPs	0 ~ 5.0 mU/mL	AChE could hydrolyze ATC to generate thiocholine, which could induces the aggregation of AuNPs	[103]
DNA methyltransferase/glycosylase	Au NPs	2–104 U/mL for Dmnt1; 1.6–256 U/mL for hOGG1	Covalent capture of target enzymes by activity-based DNA probes which created terminal protection of the DNA probes tethered on AuNPs from degradation by Exo I and III	[104]
Cells				
Ramos cell line	Au NPs	90 ~ 4,000 cells	Au NP-based colorimetric assay for the direct detection of cancerous cells	[105]
Breast cancer SK-BR-3 cell lines	Oval-shaped gold nanoparticle	$10 \sim 10^6$ cells/mL	Multifunctional (monoclonal anti-HER2/c-erb-2 antibody and S6 RNA aptamer-conjugated) oval-shaped gold-nanoparticle-based nanoconjugate	[106]

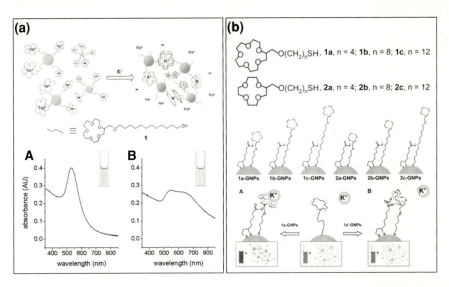

Fig. 2.14 a Schematic representation of the K^+-Induced aggregation via sandwich complexation of crown-thiol molecule 1 in a sodium-containing solution. UV-visible spectra and photographs of solutions of 2.5 mM Na^+ + 7.1 nM colloidal gold containing (A) Li^+, Cs^+, NH_4^+, Mg^{2+}, and Ca^{2+}(0.1 mM each) and (B) the above species and 0.1 mM K^+. Reprinted with permission from Ref. [67]. Copyright 2002, American Chemical Society. **b** Structures of crown thiols employed in the modification of Au NPs; Relative positions of the carboxylate and crown moiety for crown thiols with different chain lengths; Proposed structures of the crown moiety preorganized due to the neighboring molecules of (A) thioctic acid and (B) thioctic amine. The cuvettes contain 16.5 nM concentrations of the corresponding Au NPs, 100 µM K^+, and 2.5 mM Na^+. Reprinted with permission from Ref. [68]. Copyright 2005, American Chemical Society

Colorimetric detection for Alkali and alkaline earth metal ions has been investigated using Au NP-based probes. 15-crown-5 moieties functionalized Au NPs have been fabricated for the colorimetric detection of potassium ions (K^+) via formation of a 2:1 sandwich complex between 15-crown-5 moiety and K^+ [67]. This nanoprobe showed highly sensitive and selective colorimetric response toward K^+ even in the presence of physiologically important cations, including Li^+, Cs^+, NH_4^+, Mg^{2+}, Ca^{2+}, and excess Na^+ (Fig. 2.14a). Later, the performance of this probe system was improved by bi-ligand co-functionalized Au NPs with thioctic acid and crown thiols [68]. The rate of K^+ recognition increased from this system has been attributed to a cooperative effect that allows crown moiety to be preorganized by the negatively charged carboxylate moiety of the thioctic acid for K^+ recognition. Utilizing this principle the analogous detection of Na^+ in urine has been achieved by incorporating 12-crown-4 onto the Au NP surface together with the thioctic acid (Fig. 2.14b). In a similar fashion Ca^{2+} has been detected by utilizing 1-thiohexyl carboxylic acid and 1-thiohexyl β-D-lactopyranoside co-functionalized Au NPs [70].

Heavy metal ions such as Hg^{2+}, Pb^{2+}, As^{3+}, and Cd^{2+} have been well-known significant health hazards. Colorimetric nanoprobes for heavy metal ions detection have been designed using novel metal nanoparticles carrying specific ligands, such as aliphatic acid, aptamer, DNA enzyme. Hupp et al. have reported a simple colorimetric technique for the detection of trace toxic metals such as Pb^{2+}, Cd^{2+}, Hg^{2+} [107]. In this system, the aggregation of 13 nm Au NPs capped with 11-mercaptoundecanoic acid was driven by ion recognition and binding, the color change could be employed for visual sensing of the ions. Chang et al. developed a similar method for Hg^{2+} detection using mercaptopropionic acid-modified Au NPs in the presence of 2, 6-pyridinedicarboxylic acid [47]. For colorimetric Hg^{2+} detection, the specific affinity between thymine (T) and Hg^{2+} has been a considerable recognition moiety. Dependent on $T–Hg^{2+}–T$ coordination chemistry, the colorimetric assay for Hg^{2+} was first developed by the Mirkin's group based on a complementary DNA-Au NP system with designed T–T mismatches [46]. As shown in Fig. 2.15A, two types of nanoparticles aggregated together after the specially designed ssDNA hybridized, causing a red-to-purple color change. After raising the temperature to the melting temperature (T_m), the dsDNA de-hybridized, which made Au NP aggregates dissociate reversibly along with a color change back to red. Among environmentally relevant metal ions, only Hg^{2+} could raise the T_m obviously as shown in Fig. 2.15B. The sharp melting transition enhanced the sensitivity and lowered the LOD visibly to 100 nM (20 ppb) in contrast to organic colorimetric system. Based on the above works, Liu et al. developed a system that was not only selective and sensitive but also practical and convenient for colorimetric detection of Hg^{2+} at room temperature [48]. Introduction of Hg^{2+} into an aqueous solution containing oligonucleotide-tethered gold nanoparticle probes and a linker oligonucleotide with a number of T–T mismatches results in the formation of particle aggregates at room temperature with a concomitant colorimetric response. The high selectivity of this detection system is attributed to Hg^{2+}-mediated formation of $T–Hg^{2+}–T$ base pairs as evidenced by an increase in a sharp melting temperature. This colorimetric assay showed high selectivity toward Hg^{2+} in the presence of other metal ions (Fig. 2.15D).

Thymine molecule derivatives such as thiolated thymine and deoxythymidine triphosphates also have been used as the specific ligands for Hg^{2+} detection in Au NP-based colorimetric probes [30, 50, 51]. Compared with T rich-DNA oligomers, small thymine molecule derivatives seem to be more simple and easier to be combined to the surface of nanoparticles. Yang's group designed a new approach for simple and rapid colorimetric detection of Hg^{2+} based on Hg^{2+} induced aggregation of deoxythymidine triphosphates (dTTPs)-functionalized Au NPs [50]. The dTTPs could protect bare Au NPs from aggregation in the presence of salt ($KClO_4$). With the addition of Hg^{2+}, the dTTPs were desorbed from the Au NPs surface by the interaction of Hg^{2+} with the adsorbed dTTPs by forming $T–Hg^{2+}–T$ complex, thus induced aggregation of Au NPs, resulting in a red to blue–gray color change of Au NPs owing to the interparticle coupled plasmon excitons in the aggregated states (Fig. 2.15E). This method exhibited high sensitivity toward Hg^{2+} with a LOD of 50 nM (Fig. 2.15F). Later, Lou et al. proposed a colorimetric detection method for Hg^{2+} by using anti-aggregation of Au NPs based on the coordination between thymine and Hg^{2+} [51]. In this study, the thymine can

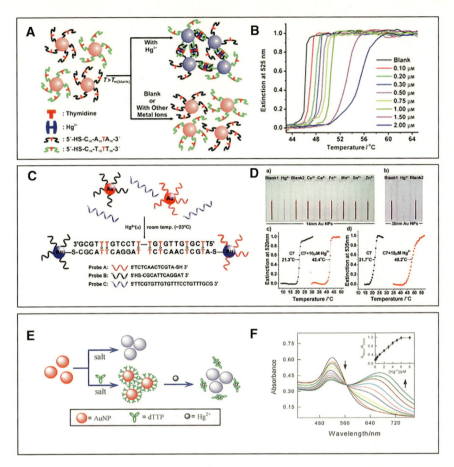

Fig. 2.15 **A** Colorimetric detection of Hg^{2+} using DNA-Au NPs and **B** Normalized melting *curves* of aggregates (probes A and B) with different concentrations of Hg^{2+}. Reproduced with permission from Ref. [46]. Copyright 2007, American Chemical Society. **C** Schematic representation of colorimetric detection of Hg^{2+} using ssDNA-Au NPs and **D** (a) Color response of a 14 nm NP detection system (probes A, B, and C7) in the presence of a selection of metal ions (Hg^{2+}, Cu^{2+}, Ca^{2+}, Fe^{3+}, Mn^{2+}, Sn^{2+}, Zn^{2+}; 10 μM each). Note that Blank1 (probes A, B, and C7 without Hg^{2+}) and Blank2 (probes A and B with Hg^{2+}) were used as control references. (b) Color response of a 30 nm NP detection system under the same conditions. (c, d) Normalized melting *curves* of the solution (containing probe A, B, and C7) with or without Hg^{2+} (10 μM) for the 14 and 30 nm NP systems, respectively. Reproduced with permission from Ref. [48]. Copyright 2008, American Chemical Society. **E** Schematic description of colorimetric sensing of Hg^{2+} based on the dTTPs-stabilized Au NPs absorption spectra changes of T–Au NPs solution in the presence of different concentrations of Hg^{2+}. **F** The *arrows* indicate the signal changes as increases in Hg^{2+} concentrations. *Inset* plot of the value of absorbance ratios, $A_{680\ nm}/A_{520\ nm}$, of T–Au NPs as a function of the concentration of Hg^{2+}. Reproduced with permission from Ref. [50]. Copyright 2011, Royal Society of Chemistry

bind to the Au NPs through Au–N bonds and induce aggregation of Au NPs. In the presence of Hg^{2+}, the thymine was released from the surface of Au NPs via the formation of a thymine-Hg^{2+} coordination complex, leading to the dispersion of Au NPs.

Colorimetric nanoprobes for Pb^{2+} detection have been largely developed by using the size and distance dependent optical properties of nanoparticles [39, 108, 109]. Huang's group designed a colorimetric assay for Pb^{2+} based on the leaching of Au NPs owing to the fact that Pb^{2+} accelerates the leaching rate of Au NPs by thiosulfate ($S_2O_3^{2-}$) and 2-mercaptoethanol (2-ME) [39]. The formation of Pb–Au alloys on the surfaces of the Au NPs in the presence of Pb^{2+} and 2-ME. The formation of Pb–Au alloys accelerated the Au NPs rapidly dissolved into solution, leading to dramatic decreases in the SPR absorption (Fig. 2.16A). The 2-ME/ $S_2O_3^{2-}$–Au NP probe is highly sensitive (LOD = 0.5 nM) and selective (by at least 1,000-fold over other metal ions) toward Pb^{2+}, with a linear detection range (2.5 nM–10 μM) over nearly four orders of magnitude.

DNAzymes, a type of DNA-based catalysts, obtained through the combinatorial method systematic evolution of ligands by exponential enrichment (SELEX), can direct assembly of Au NPs. Lu's group have fabricated highly selective Pb^{2+} sensors using DNAzyme-directed assembly of Au NPs [56]. The design of label-free DNAzyme colorimetric Pb^{2+} sensor is based on the 8–17 DNAzyme highly specific for Pb^{2+} composed of a substrate strand extended by eight bases at the 5' end (called (8)17S) and an enzyme strand extended by eight complimentary bases at 3' end (called 17E(8)) (Fig. 2.16B). The 8-base-pair extension allows stable hybridization between the substrate and enzyme strands at ambient temperature, while still allowing release of single stranded DNA (ssDNA) at the other end upon cleavage in the presence of Pb^{2+}. Upon addition of trishydroxymethylaminome-thane (Tris) and NaCl to adjust ionic strength, followed by addition of Au NPs, the released ssDNA can be adsorbed onto Au NP and prevent the individual Au NP from forming blue aggregates under high-ionic-strength conditions. In the absence of Pb^{2+} or in the presence of other metal ions, however, no cleavage reaction should occur, and the enzyme–substrate complex can not stabilize Au NPs, resulting in Au NP aggregates. This sensor showed high sensitivity with a detection limit of 3 nM. For simplicity, Su et al. developed a facile, cost-effective colorimetric detection method for Pb^{2+} by using glutathione functionalized Au NPs [57]. The modified Au NPs could be induced to aggregate immediately in the presence of Pb^{2+} with the addition of NaCl (Fig. 2.16C).

Glutathione, dithiothreitol, and cysteine co-functionalized Au NPs have been fabricated for As^{3+} detection (1 ppb) in groundwater [58]. Au NPs modified with 5, 5'-dithiobis (2-nitrobenzoic acid) were also used for the detection of trace levels Cr^{3+} (99.6 ppb) in the presence of 15 other metal ions in aqueous solution [60]. 6-mercaptonicotinic acid and L-cysteine co-functionalized Au NPs were used for colorimetric detection of Cd^{2+} (LOD = 100 nM) [59]. Generally, there is no specific recognition ligand for these metal ions, two or more ligands are used to co-modify Au NPs with synergistic effect in colorimetric assay.

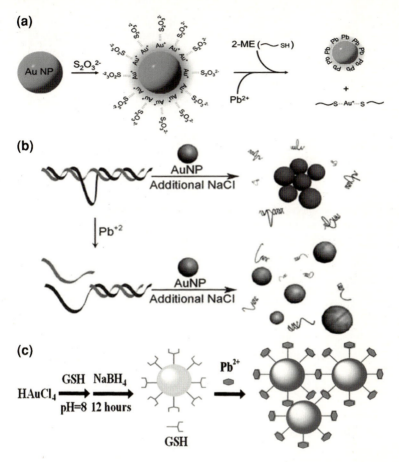

Fig. 2.16 **A** Cartoon representation of the sensing mechanism of the 2-ME/$S_2O_3^{2-}$–Au NP Probe for the detection of Pb^{2+} ions. Reproduced with permission from Ref. [39], Copyright 2009, American Chemical Society. **B** *Left* Secondary structure of the DNAzyme complex, which consists of an enzyme strand (17E(8)) and a substrate strand ((8)17S). After lead-induced cleavage, 10-mer ssDNA is released which can adsorb onto the AuNP surface. *Right* Schematic of the label-free colorimetric sensor. The lead-treated/-untreated complex and NaCl are mixed with AuNPs. The AuNPs aggregate in the absence of lead but remain dispersed in the presence of lead. Reproduced with permission from Ref. [56], Copyright 2008, Wiley. **C** Strategy for the colorimetric detection of Pb^{2+} using GSH-Au NPs Reproduced with permission from Ref. [57], Copyright 2010, American Chemical Society

Hutchison and coworkers have developed a trivalent lanthanide (Ln^{3+}) ion sensor based on tetramethylmalonamide (TMMA) functionalized Au NPs [71]. The presence of Ln^{3+} ions in the Au NP solution initiates Au NP cross-linking and concomitant red to blue color change through the formation of 2:1 TMMA-Ln^{3+} chelating complex. An immediate colorimetric response to the Ln^{3+} ions was detected, with sensitivity down to ~ 50 nM for Eu^{3+} and Sm^{3+}.

Colorimetric assay also has been explored for anions detection. The Griess reaction where sulfanilamide and naphthylethylenediamine are coupled by nitrite has been used to create colorimetric assay for NO_2^- detection. Mirkin's group designed a colorimetric nitrite sensor with Au NP probes [72]. In the system, two types of Au NPs were used, aniline Au NPs were modified with 5—[1, 2] dithiolan-3-yl-pentanoic acid [2-(4-amino-phenyl)ethyl]amide and co functionalized with hydrophobic (11-mercapto-undecyl)-trimethyl-ammonium (MTA) with molecules to increase their solubility; naphthalene Au NPs were modified with 5— [1, 2] dithiolan-3-yl-pentanoic acid [2-(naphthalene-1-ylamino)et-hyl]amide and MTA. The aniline and naphthalene Au NPs are red when dispersed in aqueous solution. In the presence of NO_2^- under acidic conditions, however, the amine groups on the aniline Au NPs are converted to a diazonium salt, which then couples with the naphthalene Au NPs to form covalently interconnected nanoparticle probes. This reaction causes the formation of crosslinked particle networks which precipitate rapidly, causing the solution to change from red to colorless (Fig. 2.17a). A novel noncrosslinking NO_2^- sensor was developed utilizing 4-aminothiophenol (4-ATP) modified Au NRs [73]. In the presence of nitrite ions, the deamination reaction was induced by heating the 4-ATP modified Au NRs in ethanol solution, resulting in the reduction of the Au NRs surface charge, which led to aggregation and a colorimetric response that was quantitatively correlated to the concentration of nitrite ions (Fig. 2.17b). This simple assay was rapid (≤ 10 min) and highly sensitive (<1 ppm of nitrite), and it can be used for rapid monitoring of drinking water quality. Zhang et al. developed a simple colorimetric method for sensing of nitrite as low as 4.0 μM by naked eyes [74]. This method is based on etching of Au NRs accompanied by shape changes in aspect ratios (length/width) and a visible color change from bluish green to red and then to colorless with the increase of nitrite (Fig. 2.17c). Colorimetric nanoprobes for other anions such as I, SCN^- also have been investigated with efforts [76, 77].

2.3.2 Detection of Small Organic Molecules

A variety of colorimetric detection methods for small organic molecules such as glucose, adenosine triphosphate (ATP), cocaine, 2, 4, 6-trinitrotoluene (TNT), dopamine, have been developed based on novel metal nanoparticles, as summarized in Table 2.1.

TNT is a leading example of a nitroaromatic explosive with significant detrimental effects on the environment and human health. Mao and coworkers reported a simple and sensitive method for the colorimetric visualization of TNT at picomolar levels by using Au NPs [78]. This method was based on the color change of Au NPs induced by the donor–acceptor (D–A) interaction between TNT and primary amines (Fig. 2.18A). In this system, cysteamine was used both as the primary amine and as the stabilizer for Au NPs. Cysteamine-stabilized AuNPs were well dispersed and the color was wine red. In the presence of TNT, the D–A

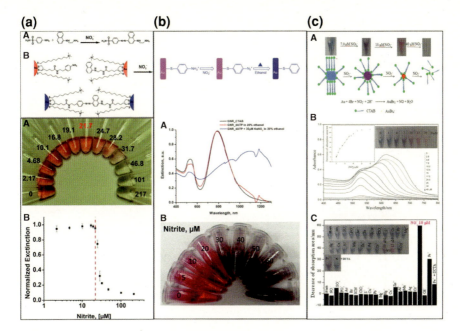

Fig. 2.17 a Griess reaction; Colorimetric detection of nitrite with functionalized Au NPs photograph of particle solutions after incubation with various concentrations of nitrite. The nitrite concentrations, in μM, are listed next to the respective solutions. The MCL of nitrite in drinking water (21.7 μM) is highlighted in *red*. (B) Particle solution extinction at 524 nm after incubation as a function of nitrite concentration. The *red dashed line* indicates the nitrite MCL. Reprinted with permission from Ref. [72], Copyright 2009, American Chemical Society. **b** Noncross-linking colorimetric detection of nitrite with 4-ATP modified Au NRs; (A) Absorption spectra of CTAB covered Au NRs (*black*), Au NR-4ATP in 20 % ethanol solution (*red*), and aggregated Au NR after heating in 20 % ethanol (*blue*). (B) Photograph of Au NR-4ATP reacted with various concentrations of nitrite after incubation in 20 % ethanol at 95 °C. Reprinted with permission from Ref. [73], Copyright 2009, American Chemical Society. **c** Schematic illustration for the colorimetric sensing of NO_2^- based on etching of Au NR; Absorption spectra of GNRs after incubation with different concentrations of NO_2^- for 20 min. *Insets* show the decrease of absorption area response to different concentrations of NO_2^- and the color change with the increase of NO_2^- concentration from *left* to *right*, respectively. Reprinted with permission from Ref. [74], Copyright 2012, Royal Society of Chemistry

interaction between TNT and cysteamine at the surface of Au NP led to the aggregation of Au NPs and the color changed to violet blue, that can be readily seen by the naked eye. The strong D–A interaction between TNT and cysteamine and the good analytical properties of Au NPs described in previous reports substantially enable a picomolar amount of TNT (0.5 pM ∼ 5 nM) to be visualized by the naked eye (Fig. 2.18B). This study essentially offers a new and simple but sensitive method for TNT detection.

Dopamine (DA) plays a central role in brain functions such as reward-related behavior, movement, and mood. Shi's group first reported a direct, selective and

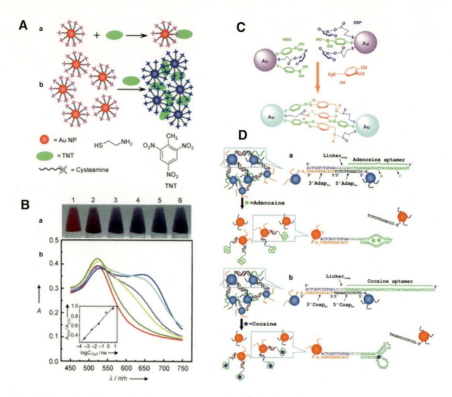

Fig. 2.18 **A** D–A interaction between cysteamine and TNT. (b) Assay for direct colorimetric visualization of TNT based on the electron **D–A** interaction at the Au NP/solution interface. **B** Colorimetric visualization of TNT by using Au NPs (containing 500 nm cysteamine). TNT concentrations varied from 5×10^{-13} M (2) to 5×10^{-9} M (6). (b) UV/Vis spectra of the Au NPs suspension (10 nm) containing 500 nm cysteamine in the presence of different concentrations of TNT: *red*, 0 m; *dark yellow*, 5×10^{-13} M; *yellow*, 5×10^{-12} M; *magenta*, 5×10^{-11} M; *cyan*, 5×10^{-10} M; *blue*, 5×10^{-9} M. *Inset* plot of A_{650}/A_{520} against log C_{TNT} for TNT assay. Reprinted with permission from Ref. [78]. Copyright 2008, Wiley. **C** Colorimetric detection of dopamine using functionalized gold nanoparticles (the MBA-DSP-AuNPs probe). Reprinted with permission from Ref. [81]. Copyright 2011, Wiley. **D** (a) Schematic representation of colorimetric detection of adenosine. The DNA sequences are shown in the *right side* of the figure. The A12 in 3'Adap Au denotes a 12-mer polyadenine chain. In a control experiment, a mutated linker with the two mutations shown by the two short *black arrows* was used. Note: The drawing is not to scale. (b) Schematic representation of the colorimetric detection of cocaine based on cocaine-induced disassembly of nanoparticle aggregates linked by a cocaine aptamer. Reprinted with permission from Ref. [87]. Copyright 2005, Wiley

sensitive strategy for the colorimetric visualization of cerebral DA at nanomolar levels using 4-mercaptophenylboronic acid (MBA) and dithiobis(succinimidyl-propionate) (DSP) cofunctionalized Au NPs [81]. MBA and DSP not only act as stabilizers for Au NPs but also interact with diols and amine functional groups, respectively, in DA to doubly recognize DA with high specificity (Fig. 2.18C). Double interactions between the functionalized Au NPs and DA triggered Au NP aggregation, thus resulting in a color change from wine red to blue and to direct colorimetric visualization of DA.

Lu reported a general method to construct sensors based on a color change of Au NPs for any aptamer of interest [87]. The sensor was made of nanoparticle aggregates containing three components: two kinds of ssDNA-modified Au NPs and a linker DNA molecule that carries adenosine aptamer (Fig. 2.18D–a). Initially, Au NPs and the linker DNA were suspended in solution to generate purple Au NPs. In the Au NP aggregation process, the linker DNA molecule pairs respectively with two ssDNA-functionalized Au NPs where a part of adenosine aptamer also takes part in the DNA hybridization process. With the presence of adenosine, the aptamer changes its structure to bind with adenosine, resulting in the disassembly of the AuNP aggregates with a concomitant blue-to-red color change. Utilizing this system, adenosine was detected in concentrations from 0.3 to 2 mM. Similarly, a colorimetric sensor for cocaine was further constructed using a cocaine aptamer (Fig. 2.18D–b).

Chiral recognition is among the important and special modes of molecular recognition. Ye et al. developed an enantioseparation and detection platform for D- and L-cysteine using uridine 5'-triphosphate (UTP)-capped silver nanoparticles [96]. As seen in Fig. 2.19-1, in the presence of D-Cys, an appreciable yellow-to-red color shift of UTP-capped AgNPs can be observed, while no color changes were found in the presence of L-Cys. The limit of discrimination concentration between L- and D-Cys is approximately 100 nM (Fig. 2.19-2). UTP-capped AgNPs selectively interacted with one enantiomer of cysteine from a solution of racemic cysteine, leaving an excess of the other enantiomer in the solution after centrifugation treatment, thus resulting in enantioselective separation (Fig. 2.19-3).

2.3.3 Detection of Oligonucleotides

Nucleic acid-based detection has attracted great interest for early diagnosis of many diseases including cancer. Nanoparticle-based colorimetric assays have been demonstrated to be a highly competitive technology for oligonucleotides targets on the basis of highly specific base-paring of DNA strands [33]. Au NPs aggregate with concomitant color change is trigged by the presence of target oligonucleotides as a result of hybridization of the DNA strand. Fabrication of Au NPs functionalized with thiolated DNA strand allowed researchers to tailor the properties of the nanoparticle probes according to the assay method. This discovery has stimulated extensive use of oligonucleotide-directed Au NP aggregation for colorimetric detection of oligonucleotides. Recently, Gao and coworkers reported a method to conduct the detection of DNA target under extremely low salt conditions where the secondary structures are less stable and more accessible [98]. In this approach, a new type of nanoparticle probes prepared by functionalizing gold nanoparticles with nonionic morpholino oligos is employed. Because of the salt-

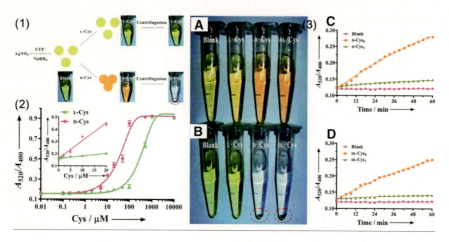

Fig. 2.19 **1** Colorimetric discrimination of L-and D-Cys using UTP-capped AgNPs. **2** Plots of A_{520}/A_{400} ratio of UTP-capped AgNPs upon the addition of l- and d-Cys at different concentrations (0.01, 0.1, 0.2, 0.5, 1, 2, 5, 10, 20, 50, 100, 500, 1,000, and 10,000 μM). *Inset* magnification of the plots in the range of 0.0–20 μM. The reaction time was 60 min, and then, the A_{520}/A_{400} ratios were collected. **3** Photo exhibition of UTP-capped AgNPs toward 100 μM l-Cys, d-Cys, and dl-Cys (A) before centrifugation and (B) after centrifugation. Plots of absorption ratios (A_{520}/A_{400}) corresponding to (C) 10 μM d-Cys0 and d-Cys1 (the supernatant of 100 μM d-Cys0 reacted with UTP-capped AgNPs) and (D) 10 μM dl-Cys0 and dl-Cys1 (the supernatant of 100 μM dl-Cys0 reacted with UTP-capped AgNPs). Reproduced with permission from Ref. [96]. Copyright 2011, American Chemical Society

independent hybridization of the probes with nucleic acid targets, nanoparticle assemblies can be formed in 2 mM Tris buffer solutions containing 0–5 mM NaCl, leading to the colorimetric target recognition (Fig. 2.20A). In this study, sharp melting transitions were observed in this method when a small amount of NaCl was presented. The melting behavior enables the unambiguous discrimination of the sequences with single-base substitution, deletion, or insertion [98]. Su and coworkers developed a label-free homogeneous phase bioassay to characterize the DNA binding properties of single-stranded DNA binding (SSB) protein using unmodified Au NPs and its application for detection of single nucleotide polymorphisms [97]. As shown in Fig. 2.20B, mismatched ssDNA can bind to SSB and form a SSB-ssDNA complex, which can protect Au NPs from salt-induced aggregation, while the complementary DNA can not, the Au NPs aggregated in the presence of salt (KCl). In this study, the detection of DNA hybridization with single nucleotide polymorphism selectivity was further developed. Owing to the high affinity between SSB and dissociated ssDNA, single-base mismatch discrimination in a long sequence of 30-mer DNA was achieved for both the end- and center-base mismatch.

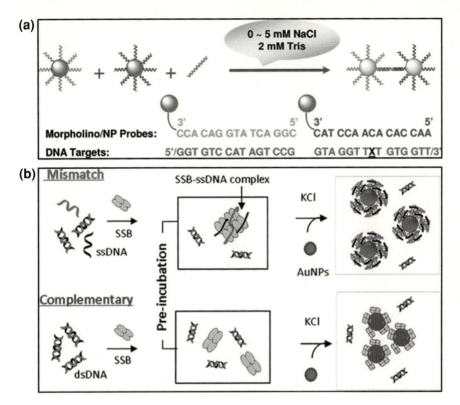

Fig. 2.20 A Schematic presentation of the colorimetric detection of nucleic acids under extremely low salt conditions. Reprinted with permission from Ref. [98]. Copyright 2011, American Chemical Society. **B** Schematic illustrations of DNA detection principle based on the inverse relationship between sequence-dependent DNA hybridization efficiency and the tendency of forming large SSB–ssDNA complex to protect AuNPs from salt-induced aggregation. Reprinted with permission from Ref. [97]. Copyright 2011, American Chemical Society

2.3.4 Detection of Proteins

Colorimetric nanoprobes have been extensively developed for the detection of disease associated biomarker proteins or irregular proteins, such as platelet-derived growth factors (PDGFs) and their receptors, thrombin, histone-modifying enzymes, as summarized in Table 2.1.

Chang and coworkers developed a highly specific sensing system for PDGFs and platelet-derived growth factor receptors (PDGFR) using aptamer-modified Au NPs [101]. Au NPs modified with an aptamer (Apt-AuNPs) that is specific to PDGFs was used in this study. The Apt-Au NP solutions change color in the presence of PDGF at high enough concentration, where PDGF molecules act as bridges that link Apt-AuNPs together. At very high PDGF concentrations, there was no obvious aggregation owing to the repulsion and steric effects because the

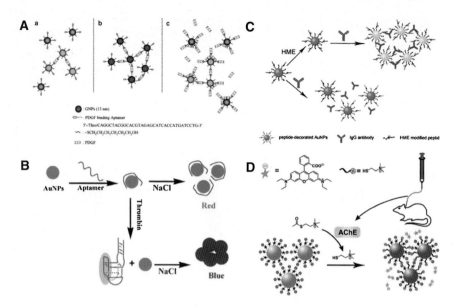

Fig. 2.21 **a** Schematic representation of the aggregation of Apt-Au NPs in the presence of PDGFs at Low, (B) Medium, and (C) High Concentrations. Reproduced with permission from Ref. [101]. Copyright 2005, American Chemical Society. **b** AuNPs colorimetric strategy for thrombin detection. Reproduced with permission from Ref. [99]. Copyright 2007, Royal Society of Chemistry. **c** Illustration of biosensing strategy for Histone-Modifying Enzyme (HME) based on antibody-mediated assembly of Au NPs decorated with substrate peptides subjected to enzymatic modifications. Reproduced with permission from Ref. [100]. Copyright 2012, American Chemical Society. **d** The detection (colorimetric and fluorometric) of AChE based on RB-Au NPs. The well-dispersed RB-Au NPs (*red*) are induced to aggregate (*purple*) via electrostatic interaction in the presence of thiocholine derived from the hydrolysis of ATC catalyzed by AChE in the CSF of transgenic mice, accompanied with the fluorescence recovery of RB (the color of the stars changed from *gray* to *green*). Reproduced with permission from Ref. [103]. Copyright 2012, Wiley

surface of the Au NPs became saturated with PDGF molecules through aptamer-PDGF binding, as shown in Fig. 2.21a. Utilizing this system, the detection limit of PDGFs was as low as 2.5 and 3.2 nM for PDGF receptor.

Dong's group reported aptamer-based colorimetric sensing of alpha-thrombin using unmodified Au NPs [99]. Thrombin binding aptamer (TBA) is much more inclined to fold into a structure of G-quadruplex/duplex when interacts with thrombin. In the absence of thrombin, unfold TBA could protect Au NPs from salt-induced aggregation and the solution remained red color. With the addition of thrombin, TBA interacted with thrombin and folded into a structure of G-quadruplex/duplex, resulting in the aggregation of AuNPs after the addition of NaCl as shown in Fig. 2.21b. The color of AuNPs colloidal solution changed from red to purple. The change of TBA conformation from the unfolded one to G-quadruplex/duplex could be directly observed with the naked eye, realizing the detection of protein thrombin in a very convenient way from 0 to 167 nM with a detection limit of 0.83 nM.

Enzymatic immuno-assembly of Au NPs for visualized activity screening of histone-modifying enzymes (HME) has been developed by Liu and coworkers [100]. This strategy relies on the antibody-mediated assembly of AuNPs decorated with substrate peptides that are subjected to enzymatic modifications by the HME Fig. 2.21c. The Au NP was decorated with a substrate peptide containing a substrate sequence at the N' terminal and a thiolated, negatively charged spacer sequence at the C' terminal. In the presence of an active HME such as HMT or HAT, the substrate peptide is modified at a specified site with a certain group such as methyl for HMT or acetyl for HAT. After the addition of an immunoglobin G (IgG) antibody specific to this modified peptide (a methylated peptide sequence for HMT or an acetylated peptide sequence for HAT), the peptides subjected to enzymatic modifications can be bound by the divalent IgG antibody at its two binding sites. This triggers a network-like assembly of the peptide-modified AuNP and thus induces a significant variation in the plasmon resonance absorption peak with a visualized color change. A quasilinear correlation was obtained to the logarithmic concentration of HMT ranging from 1 to 200 nM with a detection limit of 0.2, 50 nM to 20 μM for inhibitors of HMT, 2–200 nM with the detection limit estimated to be 0.5 nM for HAT.

Jiang's group provided a highly sensitive and selective rhodamine B-modified gold nanoparticle (RB-Au NP)-based assay with dual readouts (colorimetric and fluorometric) for monitoring the levels of acetylcholinesterase (AChE) in the cerebrospinal fluid (CSF) of transgenic mice suffering from Alzheimer's disease (AD) [103]. In this study, upon the addition of both acetylthilcholine (ATC, an analog of acetylcholine) and AChE into a RB-Au NPs solution, AChE could hydrolyze ATC to generate thiocholine, which strongly binded onto surfaces of Au NPs via the formation of Au–S bond to replace RB molecules, resulting in the desorption of RB molecules from Au surfaces. Thiocholine and the residual RB molecules attached to different Au NP surfaces may be able to interact via electrostatic interaction between the quaternary ammonium group on thiocholine and the acidic group on RB and cause the aggregation of Au NPs, as shown in Fig. 2.21d. This process resulted in a rapid change of the absorption band as well as the color change of the AuNPs solution from red to blue. In addition, when AChE inhibitors were present, AChE failed to catalyze ATC to generate the thiocholine thus the RB-AuNPs remained dispersed. For the colorimetric response, the detection limit can reach 1.0 mU/mL.

2.3.5 Sensing of Cancer Cells

The key to the effective and ultimately successful treatment of diseases such as cancer is early and accurate diagnosis. Colorimetric sensing of cancer cell does act as an attractive method based on the aptamer and specific antibody-conjugated nanoparticles. The aptamers are selected using the cell-SELEX methodology in which live whole cells served as the target. Tan and coworkers have developed a

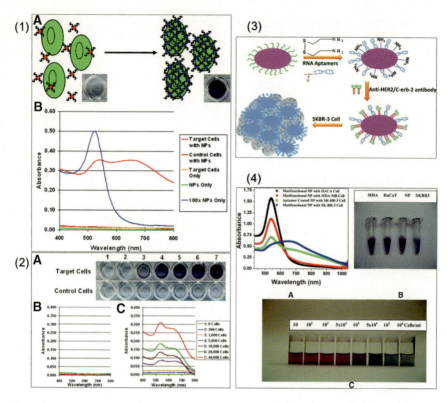

Fig. 2.22 **1** (A) Schematic representation of the aptamer-conjugated Au NP based colorimetric assay. (B) Plots depicting the absorption spectra obtained for various samples analyzed using aptamer-conjugated Au NPs. The spectra illustrate the differences in spectral characteristics observed after the aptamer-conjugated Au NPs bind to the target cells. **2** Images of aptamer-conjugated Au NPs with increasing amounts of target (*top*) and control cells (*bottom*). The amount of cells used in each sample is given in the legend on the bottom right. (B) Absorption spectra of the control cell samples with aptamer-conjugated Au NPs. (C) Absorption spectra of the target samples with aptamer-conjugated Au NPs. Reproduced with permission from Ref. [105]. Copyright 2008, American Chemical Society. **3** First two steps show schematic representation of the synthesis of monoclonal anti-HER2 antibody and S6 RNA aptamer-conjugated *oval-shaped* gold nanoparticles. Third step shows schematic representation of multifunctional *oval-shaped* gold-nanoparticle-based sensing of the SK-BR-3 breast cancer cell line. **4** (A) Absorption profile variation of multifunctional *oval-shaped* gold nanoparticles due to the addition of different cancerous and noncancerous cells. (B) Photograph showing colorimetric change upon addition of different cancer cells (10^4 cells/mL). (C) Photograph demonstrating colorimetric change upon the addition of different numbers of SK-BR-3 cells. Reproduced with permission from Ref. [106]. Copyright 2010, American Chemical Society

colorimetric assay for the direct detection of diseased cells [105]. In this study, the aptamer-conjugated Au NPs (20 nm) were targeted to assemble on the surface of a specific type of cancer cell through the recognition of the aptamer to its target on the cell membrane surface, inducing a distinct color change as shown in Fig. 2.22-

1A. The assembly of Au NPs around the target cells caused an increase in the absorption and scattering of the solution Fig. 2.22-1B. This colorimetric assay can realize the quantitative analysis of target cells (Fig. 2.22-2) and showed excellent sensitivity with both the naked eye and based on absorbance measurements with a detection limit calculated to be 90 cells.

In addition to the aptamer-conjugated nanoparticles, antibody-functionalized nanoparticles also can act as effective recognition moiety for the colorimetric selective sensing of cancer cells. Ray group reported simple colorimetric assay for breast cancer SK-BR-3 cell lines using a multifunctional (monoclonal anti-HER2/c-erb-2 antibody and S6 RNA aptamer-conjugated) oval-shaped Au NP-based nanoconjugate [106]. As shown in Fig. 2.22-3, the functionalized nanoparticles were initially fabricated through two steps; then in the presence of the breast cancer SK-BR-3 cell line, several nanoparticles can bind to HER2 receptors in the cancer cell, producing nanoparticle aggregates. Colorimetric change was observed after the addition of cancer cells as seen from Fig. 2.22-4. The use of antibody and aptamer cofunctionalized nanoconjugate platform fabricated a highly selective and sensitive detection method for a breast cancer cell line at a 100 cells/mL level.

References

1. Kelly KL, Coronado E, Zhao LL, Schatz GC (2003) The optical properties of metal nanoparticles: the influence of size, shape, and dielectric environment. J Phys Chem B 107:668–677
2. Ray PC (2010) Size and shape dependent second order nonlinear optical properties of nanomaterials and their application in biological and chemical sensing. Chem Rev 110:5332–5365
3. Halas NJ, Lal S, Chang WS, Link S, Nordlander P (2011) Plasmons in strongly coupled metallic nanostructures. Chem Rev 111:3913–3961
4. Jain PK, Huang X, El-Sayed IH, El-Sayed MA (2008) Noble metals on the nanoscale: optical and photothermal properties and some applications in imaging, sensing, biology, and medicine. Acc Chem Res 41:1578–1586
5. Wang Z, Ma L (2009) Gold nanoparticle probes. Coord Chem Rev 253:1607–1618
6. Boisselier E, Astruc D (2009) Gold nanoparticles in nanomedicine: preparations, imaging, diagnostics, therapies and toxicity. Chem Soc Rev 38:1759–1782
7. Saha K, Agasti SS, Kim C, Li X, Rotello VM (2012) Gold nanoparticles in chemical and biological sensing. Chem Rev 112:2739–2779
8. Wilson R (2008) The use of gold nanoparticles in diagnostics and detection. Chem Soc Rev 37:2028–2045
9. Motl NE, Smith AF, DeSantisa CJ, Skrabalak SE (2014) Engineering plasmonic metal colloids through composition and structural design. Chem Soc Rev. doi:10.1039/C3CS60347D
10. Zhang JZ, Noguez C (2008) Plasmonic optical properties and applications of metal nanostructures. Plasmonics 3:127–150
11. Haes AJ, Haynes CL, McFarland AD, Schatz GC, Van Duyne RP, Zou S (2005) Plasmonic materials for surface-enhanced sensing and spectroscopy. MRS Bull 30:368–375

12. Jain PK, Lee KS, El-Sayed IH, El-Sayed MA (2006) Calculated absorption and scattering properties of gold nanoparticles of different size, shape, and composition: applications in biological imaging and biomedicine. J Phys Chem B 110:7238–7248
13. Link S, El-Sayed MA (2003) Optical properties and ultrafast dynamics of metallic nanocrystals. Annu Rev Phys Chem 54:331–366
14. Griffin J, Singh AK, Senapati D, Lee E, Gaylor K, Jones-Boone J, Ray PC (2009) Sequence-specific HCV RNA quantification using the size-dependent nonlinear optical properties of gold nanoparticles. Small 5:839–845
15. Link S, El-Sayed MA (1999) Spectral properties and relaxation dynamics of surface plasmon electronic oscillations in gold and silver nanodots and nanorods. J Phys Chem B 103:8410–8426
16. Ye X, Jin L, Caglayan H, Chen J, Xing G, Zheng C, Doan-Nguyen V, Kang Y, Engheta N, Kagan CR (2012) Improved size-tunable synthesis of monodisperse gold nanorods through the use of aromatic additives. ACS Nano 6:2804–2817
17. Jakab A, Rosman C, Khalavka Y, Becker J, Trügler A, Hohenester U, Sönnichsen C (2011) Highly sensitive plasmonic silver nanorods. ACS Nano 5:6880–6885
18. Jones MR, Osberg KD, Macfarlane RJ, Langille MR, Mirkin CA (2011) Templated techniques for the synthesis and assembly of plasmonic nanostructures. Chem Rev 111:3736–3827
19. Murphy CJ, Thompson LB, Alkilany AM, Sisco PN, Boulos SP, Sivapalan ST, Yang JA, Chernak DJ, Huang J (2010) The many faces of gold nanorods. J Phys Chem Lett 1:2867–2875
20. Zeng J, Roberts S, Xia Y (2010) Nanocrystal-based time–temperature indicators. Chem Eur J 16:12559–12563
21. Singh AK, Senapati D, Neely A, Kolawole G, Hawker C, Ray PC (2009) Nonlinear optical properties of triangular silver nanomaterials. Chem Phys Lett 481:94–98
22. Becker J, Zins I, Jakab A, Khalavka Y, Schubert O, Sönnichsen C (2008) Plasmonic focusing reduces ensemble linewidth of silver-coated gold nanorods. Nano Lett 8:1719–1723
23. Xiang Y, Wu X, Liu D, Li Z, Chu W, Feng L, Zhang K, Zhou W, Xie S (2008) Gold nanorod-seeded growth of silver nanostructures: from homogeneous coating to anisotropic coating. Langmuir 24:3465–3470
24. Park G, Lee C, Seo D, Song H (2012) Full-color tuning of surface plasmon resonance by compositional variation of Au@Ag core-shell nanocubes with sulfides. Langmuir 28:9003–9009
25. Su K-H, Wei Q-H, Zhang X, Mock J, Smith DR, Schultz S (2003) Interparticle coupling effects on plasmon resonances of nanogold particles. Nano Lett 3:1087–1090
26. Daniel M-C, Astruc D (2004) Gold nanoparticles: assembly, supramolecular chemistry, quantum-size-related properties, and applications toward biology, catalysis, and nanotechnology. Chem Rev 104:293–346
27. Liu M, Guyot-Sionnest P (2004) Synthesis and optical characterization of Au/Ag core/shell nanorods. J Phys Chem B 108:5882–5888
28. Ma Y, Li W, Cho EC, Li Z, Yu T, Zeng J, Xie Z, Xia Y (2010) Au@ Ag core–shell nanocubes with finely tuned and well-controlled sizes, shell thicknesses, and optical properties. ACS Nano 4:6725–6734
29. Srivastava S, Frankamp BL, Rotello VM (2005) Controlled plasmon resonance of gold nanoparticles self-assembled with PAMAM dendrimers. Chem Mater 17:487–490
30. Chen L, Lou TT, Yu CW, Kang Q, Chen LX (2011) N-1-(2-mercaptoethyl)thymine modification of gold nanoparticles: a highly selective and sensitive colorimetric chemosensor for Hg^{2+}. Analyst 136:4770–4773
31. Zhou Y, Wang S, Zhang K, Jiang X (2008) Visual detection of copper(II) by azide- and alkyne-functionalized gold nanoparticles using click chemistry. Angew Chem Int Ed 47:7454–7456

32. Cao R, Li B, Zhang Y, Zhang Z (2011) Naked-eye sensitive detection of nuclease activity using positively-charged gold nanoparticles as colorimetric probes. Chem Commun 47:12301–12303

33. Elghanian R, Storhoff JJ, Mucic RC, Letsinger RL, Mirkin CA (1997) Selective colorimetric detection of polynucleotides based on the distance-dependent optical properties of gold nanoparticles. Science 277:1078–1081

34. Ai K, Liu Y, Lu L (2009) Hydrogen-bonding recognition-induced color change of gold nanoparticles for visual detection of melamine in raw milk and infant formula. J Am Chem Soc 131:9496–9497

35. Rosi NL, Mirkin CA (2005) Nanostructures in biodiagnostics. Chem Rev 105:1547–1562

36. Li D, Wieckowska A, Willner I (2008) Optical analysis of Hg^{2+} ions by oligonucleotide–gold-nanoparticle hybrids and DNA-based machines. Angew Chem Int Ed 120:3991–3995

37. Fu XL, Chen LX, Li JH, Lin M, You HY, Wang WH (2012) Label-free colorimetric sensor for ultrasensitive detection of heparin based on color quenching of gold nanorods by graphene oxide. Biosens Bioelectron 34:227–231

38. Lou TT, Chen ZP, Wang YQ, Chen LX (2011) Blue-to-red colorimetric sensing strategy for Hg(2+) and Ag(+) via redox-regulated surface chemistry of gold nanoparticles. ACS Appl Mater Interfaces 3:1568–1573

39. Chen Y-Y, Chang H-T, Shiang Y-C, Hung Y-L, Chiang C-K, Huang C-C (2009) Colorimetric assay for lead ions based on the leaching of gold nanoparticles. Anal Chem 81:9433–9439

40. Malile B, Chen JI (2013) Morphology-based plasmonic nanoparticle sensors: controlling etching kinetics with target-responsive permeability gate. J Am Chem Soc 135:16042–16045

41. Rex M, Hernandez FE, Campiglia AD (2006) Pushing the limits of mercury sensors with gold nanorods. Anal Chem 78:445–451

42. Lou TT, Chen LX, Chen ZP, Wang YQ, Chen L, Li JH (2011) Colorimetric detection of trace copper ions based on catalytic leaching of silver-coated gold nanoparticles. ACS Appl Mater Interfaces 3:4215–4220

43. Wang XK, Chen L, Chen LX (2013) Colorimetric determination of copper ions based on the catalytic leaching of silver from the shell of silver-coated gold nanorods. Microchim Acta 181:105–110

44. Wang GQ, Chen ZP, Chen LX (2011) Mesoporous silica-coated gold nanorods: towards sensitive colorimetric sensing of ascorbic acid via target-induced silver overcoating. Nanoscale 3:1756–1759

45. Xia Y, Ye J, Tan K, Wang J, Yang G (2013) Colorimetric visualization of glucose at the submicromole level in serum by a homogenous silver nanoprism-glucose oxidase system. Anal Chem 85:6241–6247

46. Lee JS, Han MS, Mirkin CA (2007) Colorimetric detection of mercuric ion (Hg^{2+}) in aqueous media using DNA-functionalized gold nanoparticles. Angew Chem Int Ed 46:4093–4096

47. Huang CC, Chang HT (2007) Parameters for selective colorimetric sensing of mercury(II) in aqueous solutions using mercaptopropionic acid-modified gold nanoparticles. Chem Commun 12:1215–1217

48. Xue X, Wang F, Liu X (2008) One-step, room temperature, colorimetric detection of mercury (Hg^{2+}) using DNA/nanoparticle conjugates. J Am Chem Soc 130:3244–3245

49. Yu CJ, Cheng TL, Tseng WL (2009) Effects of Mn^{2+} on oligonucleotide-gold nanoparticle hybrids for colorimetric sensing of Hg^{2+}: improving colorimetric sensitivity and accelerating color change. Biosens Bioelectron 25:204–210

50. Xu Y, Deng L, Wang H, Ouyang X, Zheng J, Li J, Yang R (2011) Metal-induced aggregation of mononucleotides-stabilized gold nanoparticles: an efficient approach for simple and rapid colorimetric detection of Hg(II). Chem Commun 47:6039–6041

51. Lou T, Chen L, Zhang C, Kang Q, You H, Shen D, Chen L (2012) A simple and sensitive colorimetric method for detection of mercury ions based on anti-aggregation of gold nanoparticles. Anal Methods 4:488
52. Wang GL, Zhu XY, Jiao HJ, Dong YM, Li ZJ (2012) Ultrasensitive and dual functional colorimetric sensors for mercury (II) ions and hydrogen peroxide based on catalytic reduction property of silver nanoparticles. Biosens Bioelectron 31:337–342
53. Chen L, Fu XL, Lu WH, Chen LX (2013) Highly sensitive and selective colorimetric sensing of Hg^{2+} based on the morphology transition of silver nanoprisms. ACS Appl Mater Interfaces 5:284–290
54. Lin C-Y, Yu C-J, Lin Y-H, Tseng W-L (2010) Colorimetric sensing of silver (I) and mercury (II) ions based on an assembly of Tween 20-stabilized gold nanoparticles. Anal Chem 82:6830–6837
55. Guo Y, Wang Z, Qu W, Shao H, Jiang X (2011) Colorimetric detection of mercury, lead and copper ions simultaneously using protein-functionalized gold nanoparticles. Biosens Bioelectron 26:4064–4069
56. Wang Z, Lee JH, Lu Y (2008) Label-free colorimetric detection of lead ions with a nanomolar detection limit and tunable dynamic range by using gold nanoparticles and DNAzyme. Adv Mater 20:3263–3267
57. Chai F, Wang C, Wang T, Li L, Su Z (2010) Colorimetric detection of Pb^{2+} using glutathione functionalized gold nanoparticles. ACS Appl Mater Interfaces 2:1466–1470
58. Kalluri JR, Arbneshi T, Khan SA, Neely A, Candice P, Varisli B, Washington M, McAfee S, Robinson B, Banerjee S, Singh AK, Senapati D, Ray PC (2009) Use of gold nanoparticles in a simple colorimetric and ultrasensitive dynamic light scattering assay: selective detection of arsenic in groundwater. Angew Chem Int Ed 48:9668–9671
59. Xue Y, Zhao H, Wu Z, Li X, He Y, Yuan Z (2011) Colorimetric detection of Cd^{2+} using gold nanoparticles cofunctionalized with 6-mercaptonicotinic acid and L-cysteine. Analyst 136:3725–3730
60. Dang YQ, Li HW, Wang B, Li L, Wu Y (2009) Selective detection of trace Cr^{3+} in aqueous solution by using 5,5'-dithiobis (2-nitrobenzoic acid)-modified gold nanoparticles. ACS Appl Mater Interfaces 1:1533–1538
61. Li F-M, Liu J-M, Wang X-X, Lin L-P, Cai W-L, Lin X, Zeng Y-N, Li Z-M, Lin S-Q (2011) Non-aggregation based label free colorimetric sensor for the detection of Cr (VI) based on selective etching of gold nanorods. Sensor Actuators B Chem 155:817–822
62. Zhang Z, Zhang J, Lou T, Pan D, Chen L, Qu C, Chen Z (2012) Label-free colorimetric sensing of cobalt(II) based on inducing aggregation of thiosulfate stabilized gold nanoparticles in the presence of ethylenediamine. Analyst 137:400–405
63. Ma YR, Niu HY, Zhang XL, Cai YQ (2011) Colorimetric detection of copper ions in tap water during the synthesis of silver/dopamine nanoparticles. Chem Commun 47:12643–12645
64. Wang SS, Chen ZP, Chen L, Liu RL, Chen LX (2013) Label-free colorimetric sensing of copper(II) ions based on accelerating decomposition of H2O2 using gold nanorods as an indicator. Analyst 138:2080–2084
65. Chen ZP, Liu RL, Wang SS, Qu CL, Chen LX, Wang Z (2013) Colorimetric sensing of copper(ii) based on catalytic etching of gold nanorods. RSC Adv 3:13318
66. Hung YL, Hsiung TM, Chen YY, Huang CC (2010) A label-free colorimetric detection of lead ions by controlling the ligand shells of gold nanoparticles. Talanta 82:516–522
67. Lin S-Y, Liu S-W, Lin C-M, Chen C-H (2002) Recognition of potassium ion in water by 15-crown-5 functionalized gold nanoparticles. Anal Chem 74:330–335
68. Lin S-Y, Chen C-H, Lin M-C, Hsu H-F (2005) A cooperative effect of bifunctionalized nanoparticles on recognition: sensing alkali ions by crown and carboxylate moieties in aqueous media. Anal Chem 77:4821–4828
69. Kim S, Kim J, Lee NH, Jang HH, Han MS (2011) A colorimetric selective sensing probe for calcium ions with tunable dynamic ranges using cytidine triphosphate stabilized gold nanoparticles. Chem Commun 47:10299–10301

70. Eom MS, Jang W, Lee YS, Choi G, Kwon YU, Han MS (2012) A bi-ligand co-functionalized gold nanoparticles-based calcium ion probe and its application to the detection of calcium ions in serum. Chem Commun 48:5566–5568

71. Lisowski CE, Hutchison JE (2009) Malonamide-functionalized gold nanoparticles for selective, colorimetric sensing of trivalent lanthanide ions. Anal Chem 81:10246–10253

72. Daniel WL, Han MS, Lee J-S, Mirkin CA (2009) Colorimetric nitrite and nitrate detection with gold nanoparticle probes and kinetic end points. J Am Chem Soc 131:6362–6363

73. Xiao N, Yu C (2010) Rapid-response and highly sensitive noncross-linking colorimetric nitrite sensor using 4-aminothiophenol modified gold nanorods. Anal Chem 82:3659–3663

74. Chen ZP, Zhang ZY, Qu CL, Pan DW, Chen LX (2012) Highly sensitive label-free colorimetric sensing of nitrite based on etching of gold nanorods. Analyst 137:5197–5200

75. Tripathy SK, Woo JY, Han CS (2011) Highly selective colorimetric detection of hydrochloric acid using unlabeled gold nanoparticles and an oxidizing agent. Anal Chem 83:9206–9212

76. Chen L, Lu WH, Wang XK, Chen LX (2013) A highly selective and sensitive colorimetric sensor for iodide detection based on anti-aggregation of gold nanoparticles. Sensor Actuators B Chem 182:482–488

77. Zhang ZY, Zhang J, Qu CL, Pan DW, Chen ZP, Chen LX (2012) Label free colorimetric sensing of thiocyanate based on inducing aggregation of Tween 20-stabilized gold nanoparticles. Analyst 137:2682–2686

78. Jiang Y, Zhao H, Zhu N, Lin Y, Yu P, Mao LQ (2008) A simple assay for direct colorimetric visualization of trinitrotoluene at picomolar levels using gold nanoparticles. Angew Chem Int Ed 47:8601–8604

79. Dasary SS, Senapati D, Singh AK, Anjaneyulu Y, Yu H, Ray PC (2010) Highly sensitive and selective dynamic light-scattering assay for TNT detection using p-ATP attached gold nanoparticle. ACS Appl Mater Interfaces 2:3455–3460

80. Radhakumary C, Sreenivasan K (2011) Naked eye detection of glucose in urine using glucose oxidase immobilized gold nanoparticles. Anal Chem 83:2829–2833

81. Kong B, Zhu A, Luo Y, Tian Y, Yu Y, Shi G (2011) Sensitive and selective colorimetric visualization of cerebral dopamine based on double molecular recognition. Angew Chem Int Ed 50:1837–1840

82. Feng JJ, Guo H, Li YF, Wang YH, Chen WY, Wang AJ (2013) Single molecular functionalized gold nanoparticles for hydrogen-bonding recognition and colorimetric detection of dopamine with high sensitivity and selectivity. ACS Appl Mater Interfaces 5:1226–1231

83. Guo L, Zhong J, Wu J, Fu F, Chen G, Zheng X, Lin S (2010) Visual detection of melamine in milk products by label-free gold nanoparticles. Talanta 82:1654–1658

84. Zhang Y, Li B, Xu C (2010) Visual detection of ascorbic acid via alkyne-azide click reaction using gold nanoparticles as a colorimetric probe. Analyst 135:1579–1584

85. Wang J, Wang L, Liu X, Liang Z, Song S, Li W, Li G, Fan C (2007) A gold nanoparticle-based aptamer target binding readout for ATP assay. Adv Mater 19:3943–3946

86. Liu J, Lu Y (2004) Adenosine-dependent assembly of aptazyme-functionalized gold nanoparticles and its application as a colorimetric biosensor. Anal Chem 76:1627–1632

87. Liu J, Lu Y (2005) Fast colorimetric sensing of adenosine and cocaine based on a general sensor design involving aptamers and nanoparticles. Angew Chem Int Ed 45:90–94

88. Zhang J, Wang L, Pan D, Song S, Boey FY, Zhang H, Fan C (2008) Visual cocaine detection with gold nanoparticles and rationally engineered aptamer structures. Small 4:1196–1200

89. Sun J, Ge J, Liu W, Fan Z, Zhang H, Wang P (2011) Highly sensitive and selective colorimetric visualization of streptomycin in raw milk using Au nanoparticles supramolecular assembly. Chem Commun 47:9888–9890

90. Zhang X, Zhao H, Xue Y, Wu Z, Zhang Y, He Y, Li X, Yuan Z (2012) Colorimetric sensing of clenbuterol using gold nanoparticles in the presence of melamine. Biosens Bioelectron 34:112–117

91. Kim YS, Kim JH, Kim IA, Lee SJ, Jurng J, Gu MB (2010) A novel colorimetric aptasensor using gold nanoparticle for a highly sensitive and specific detection of oxytetracycline. Biosens Bioelectron 26:1644–1649

92. Sun J, Guo L, Bao Y, Xie J (2011) A simple, label-free AuNPs-based colorimetric ultrasensitive detection of nerve agents and highly toxic organophosphate pesticide. Biosens Bioelectron 28:152–157

93. Zhang M, Liu YQ, Ye BC (2011) Rapid and sensitive colorimetric visualization of phthalates using UTP-modified gold nanoparticles cross-linked by copper(II). Chem Commun 47:11849–11851

94. Li L, Li B (2009) Sensitive and selective detection of cysteine using gold nanoparticles as colorimetric probes. Analyst 134:1361–1365

95. Sudeep P, Joseph SS, Thomas KG (2005) Selective detection of cysteine and glutathione using gold nanorods. J Am Chem Soc 127:6516–6517

96. Zhang M, Ye BC (2011) Colorimetric chiral recognition of enantiomers using the nucleotide-capped silver nanoparticles. Anal Chem 83:1504–1509

97. Tan YN, Lee KH, Su X (2011) Study of single-stranded DNA binding protein-nucleic acids interactions using unmodified gold nanoparticles and its application for detection of single nucleotide polymorphisms. Anal Chem 83:4251–4257

98. Zu Y, Ting AL, Yi G, Gao Z (2011) Sequence-selective recognition of nucleic acids under extremely low salt conditions using nanoparticle probes. Anal Chem 83:4090–4094

99. Wei H, Li B, Li J, Wang E, Dong S (2007) Simple and sensitive aptamer-based colorimetric sensing of protein using unmodified gold nanoparticle probes. Chem Commun 36:3735–3737

100. Zhen Z, Tang LJ, Long H, Jiang JH (2012) Enzymatic immuno-assembly of gold nanoparticles for visualized activity screening of histone-modifying enzymes. Anal Chem 84:3614–3620

101. Huang C-C, Huang Y-F, Cao Z, Tan W, Chang H-T (2005) Aptamer-modified gold nanoparticles for colorimetric determination of platelet-derived growth factors and their receptors. Anal Chem 77:5735–5741

102. Xue W, Zhang G, Zhang D (2011) A sensitive colorimetric label-free assay for trypsin and inhibitor screening with gold nanoparticles. Analyst 136:3136–3141

103. Liu D, Chen W, Tian Y, He S, Zheng W, Sun J, Wang Z, Jiang X (2012) A highly sensitive gold-nanoparticle-based assay for acetylcholinesterase in cerebrospinal fluid of transgenic mice with Alzheimer's disease. Adv Healthc Mater 1:90–95

104. Wu Z, Wu ZK, Tang H, Tang LJ, Jiang JH (2013) Activity-based DNA-gold nanoparticle probe as colorimetric biosensor for DNA methyltransferase/glycosylase assay. Anal Chem 85:4376–4383

105. Medley CD, Smith JE, Tang Z, Wu Y, Bamrungsap S, Tan W (2008) Gold nanoparticle-based colorimetric assay for the direct detection of cancerous cells. Anal Chem 80:1067–1072

106. Lu W, Arumugam SR, Senapati D, Singh AK, Arbneshi T, Khan SA, Yu H, Ray PC (2010) Multifunctional oval-shaped gold-nanoparticle-based selective detection of breast cancer cells using simple colorimetric and highly sensitive two-photon scattering assay. ACS Nano 4:1739–1749

107. Kim Y, Johnson RC, Hupp JT (2001) Gold nanoparticle-based sensing of "spectroscopically silen" heavy metal ions. Nano Lett 1:165–167

108. Beqa L, Singh AK, Khan SA, Senapati D, Arumugam SR, Ray PC (2011) Gold nanoparticle-based simple colorimetric and ultrasensitive dynamic light scattering assay for the selective detection of Pb(II) from paints, plastics, and water samples. ACS Appl Mater Interfaces 3:668–673

109. Wei H, Li B, Li J, Dong S, Wang E (2008) DNAzyme-based colorimetric sensing of lead (Pb(2+)) using unmodified gold nanoparticle probes. Nanotechnology 19:095501

Chapter 3
Fluorescent Nanoprobes

Abstract With the development of nanotechnology, nanomaterials-based fluorescence sensors have been successfully fabricated for various applications. These nanomaterials include quantum dots, noble metal nanoclusters, upconversion nanoparticles, carbon nanostructures, and so on. In this chapter, the overviews of the fluorescence properties of nanomaterials, fluorescence sensing strategies, as well as their applications in biological and chemical sensing are given in detail.

Keywords Fluorescence nanoprobes · Quantum dots · Noble metal nanoclusters · Upconversion nanoparticles · Carbon nanostructures

3.1 Fluorescence Properties of Nanomaterials

Fluorescence assays based on organic dyes have been used for many years in environmental and clinical chemistry as reliable and sensitive strategies to detect low concentrations of analytes in different matrices [1–7]. In recent years, great progress has been achieved for the development of highly sensitive chemical and biological fluorescence nanosensor via diverse transduction modes taking advantage of the unique features of novel fluorescent nanomaterials including small size effects, large surface-to-volume ratios and tunable optical properties [8–13]. An explosion of research in this field has yielded many fluorescence detection systems using various kinds of nanomaterials such as quantum dots (QDs) [14], noble metal nanoparticles [15, 16], upconversion nanoparticles (UCNPs) [17], carbon nanotubes (CNTs) [18], graphene oxides (GO) [3], and carbon nanoparticles [19] (Fig. 3.1).

3.1.1 Quantum Dots

Semiconductor QDs exhibit unique luminescence properties such as narrow and symmetric emission with tunable colors, broad and strong absorption, reasonable stability, and solution processibility [20–24]. Moreover, such luminescence

L. Chen et al., *Novel Optical Nanoprobes for Chemical and Biological Analysis*,
SpringerBriefs in Molecular Science, DOI: 10.1007/978-3-662-43624-0_3,
© The Author(s) 2014

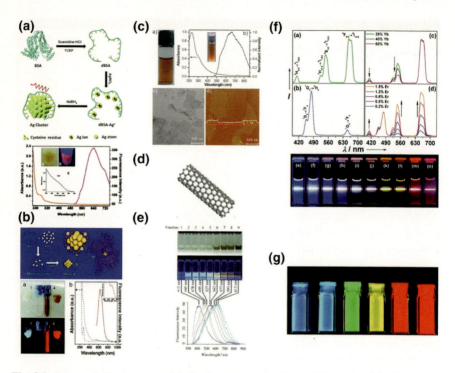

Fig. 3.1 Representative nanomaterials for fluorescence sensing. **a** Schematic of the fluorescent Ag clusters and the absorption (*yellow*) and fluorescence emission (*pink*) spectra of the as-prepared Ag clusters. Reproduced with permission from Ref. [15]. Copyright 2011, American Chemical Society. **b** Schematic of the formation of fluorescent Au clusters, the photographs of BSA (*a1, a2*) and BSA-Au NCs (*a3, a4*), and the optical absorption (*b: dash lines*) and photoemission spectra of BSA and BSA-Au NCs (*b: solid line*). The inset shows the photoexcitation spectrum of BSA-Au NCs. Reproduced with permission from Ref. [16]. Copyright 2009, American Chemical Society. **c** (*a*) Photograph for the GO colloids illuminated under sunlight; (*b*) typical absorption (*left*) and vis-NIR fluorescence emission (*right* λex = 450 nm) spectra of the GO colloids (*inset* photograph for the GO colloids excited by 365 nm); (*c*) TEM and (*d*) AFM images of the as-obtained GO nanosheets. Reproduced with permission from Ref. [3]. Copyright 2011, Royal Society of Chemistry. **d** Schematic illustration of carbon nanotubes. Reproduced with permission from Ref. [18]. Copyright 2013, Royal Society of Chemistry. **e** Optical characterization of the purified carbon nanoparticles (CNPs). Optical images illuminated under white (*top*) and UV light (312 nm; *center*). *Bottom* Fluorescence emission spectra (excitation at 315 nm) of the corresponding CNP solutions. Reproduced with permission from Ref. [19]. Copyright 2007, Wiley. **f** Room temperature upconversion emission spectra of (*a*) NaYF$_4$:Yb/Er (18/2 mol%), (*b*) NaYF$_4$:Yb/Tm (20/0.2 mol%), (*c*) NaYF$_4$:Yb/Er (25–60/2 mol%), and (*d*) NaYF$_4$:Yb/Tm/Er (20/0.2/0.2–1.5 mol%) particles in ethanol solutions (10 mM). The spectra in (*c*) and (*d*) were normalized to Er^{3+} 650 nm and Tm^{3+} 480 nm emissions, respectively. Compiled luminescent photos showing corresponding colloidal solutions of (*e*) NaYF$_4$:Yb/Tm (20/0.2 mol%), (f–j) NaYF$_4$:Yb/Tm/Er (20/0.2/0.2–1.5 mol%), and (k–n) NaYF$_4$:Yb/Er (18–60/2 mol%). Reproduced with permission from Ref. [17]. Copyright 2008, American Chemical Society. **g** Photographs of different size samples of (CdSe)ZnS. Reproduced with permission from Ref. [14]. Copyright 1997, American Chemical Society

properties can be tuned by controlling the sizes, shapes and compositions, due to the quantum confinement of the electrons in the particles. As the particles become smaller, the luminescence energies are blue-shifted to higher energies. The fluorescence spectra of the QDs are usually characterized by narrow emission bands exhibiting a large Stokes shift and broad absorbance bands. This feature allows the photo excitation of different sized QDs by a single wavelength, while generating the size controlled, different colored luminescence [25, 26].

3.1.2 Noble Metal Nanoclusters

Noble metal nanoclusters typically consist of several to tens of atoms [15, 16, 27–29]. Generally, the current methods for the preparation of noble metal nanoclusters are divided into "bottom–up" and "top–down" types [30, 31]. For a "bottom–up" approach, metal precursors are reduced to atoms and then the clusters are formed by piling metal atoms one by one [30–33]. For instance, Chang's group reported a one-pot approach to prepare fluorescent DNA-templated gold/silver nanoclusters (DNA-Au/Ag NCs) from Au^{3+}, Ag^+ and DNA (5′-CCCTTAATCCCC-3′) in the presence of $NaBH_4$ [33]. Lu et al. prepared gold clusters in NaOH solution using bovine serum albumin (BSA) as stabilization agent [34]. In contrast, for "top–down" approach, smaller metal clusters are created by core etching of metallic nanoparticles into smaller NCs by etching molecules [29, 30]. Pradeep's group prepared a bright-red-emitting new subnanocluster Au_{23} by the core etching of a widely explored and more stable cluster $Au_{25}SG_{18}$ (SG is glutathione thiolate) [35]. Nie et al. reported a highly fluorescent gold clusters by a novel ligand-induced etching process, in which hyperbranched and multivalent coordinating polymers polyethylenimine react with preformed gold nanocrystals to form atomic gold clusters [36]. The size of noble metal nanoclusters is less than 1 nm and is comparable to the Fermi wavelength of the conduction electrons. The spatial confinement of free electrons in metal nanoclusters results in discrete and size-tunable electrons transitions, leading to molecular-like properties such as luminescence and unique charging properties [37].

3.1.3 Upconversion Nanoparticles

Rare-earth UCNPs exhibit upconversion luminescence (UCL) properties upon low levels of irradiation in the near infrared (NIR) spectral region, which arising from $4f$ orbital transitions, due to the effective shielding of the $4f$ orbitals by higher lying $5s$ and $5p$ orbitals, and then minimizing the effect of the outer ligand field [13]. Generally, UCNPs comprise inorganic host, sensitizer and activator. Host materials need to have low lattice phonon energies, which is a requirement to minimize nonradiative losses and maximize the radiative emission, such as fluorides, oxides, heavy halides (chlorides, bromides and iodides), oxysulfide, phosphates, and vanadates. Among these, fluoride is the most popular host materials due to its low phonon energies (~ 350 cm^{-1}) and high chemical stability. Activators are used to generate UCL emission under NIR excitation, for example, Er^{3+}, Tm^{3+}, and Ho^{3+}, which have ladder-like arrangement energy levels. Yb^{3+}, which has a larger absorption cross-section at around 980 nm than other lanthanide ions, is often co-doped with Er^{3+}, Tm^{3+} and Ho^{3+} as a sensitizer to enhance the UCL efficiency. To date, hydro (solvo) thermal synthesis and thermal decomposition have become the two most popular methods for synthesizing of high-quality uniform UCNPs. For example, Zhang et al. reported an efficient and user-friendly method for the synthesis of uniform hexagonal-phase $NaYF_4$:Yb, Er/Tm nanocrystals with controllable shape and strong UCL emission [38]. Yan et al. developed a general synthesis of high-quality cubic (α-phase) and hexagonal (β-phase) $NaREF_4$ (RE: Pr to Lu, Y) nanocrystals (nanopolyhedra, nanorods, nanoplates, and nanospheres) and $NaYF_4$:Yb, Er/Tm nanocrystals (nanopolyhedra and nanoplates) via the co-thermolysis of $Na(CF_3COO)$ and $RE(CF_3COO)_3$ in oleic acid/oleylamine/1-octadecene [39]. With doped into suitable matrices, rare-earth nanoparticles can generate UCL from the violet to NIR region. Such upconversion mechanisms have been recognized to be involved into four main classes either alone or in combination: excited state absorption (ESA), energy transfer UC (ETU), photon avalanche (PA) and energy migration-mediated UC (EMU) (Fig. 3.2) [13, 40]. In particular, the UCL properties have the advantages of sharp emission lines, long lifetimes, a large anti-Stokes shift of several hundred nanometers, superior photostability and nonblinking, which makes UCNPs promising as bioimaging probes with attractive features, such as no auto-fluorescence from biological samples and a large penetration depth.

3.1.4 Carbon Nanostructures

Carbon nanostructures, such as CNTs, GO, and carbon dots, have been explored extensively in recent years. These carbon nanostructures have unique electrical, physical, and fluorescence properties and versatile chemistry. Based on these excellent features, carbon nanostructures have demonstrated their potential for various applications in biological and chemical sensing.

GO is expected to exhibit unique optical properties as evidenced by the recent demonstration of luminescence from GO [41–43]. This luminescence was found to occur from near UV-to-blue visible to NIR wavelength range, which originated from the recombination of electron–hole (e–h) pairs localized within small sp^2 carbon clusters embedded within an sp^3 matrix. The PL intensity varies with the nature of the reduction treatment, and can be correlated to the evolution of very small sp^2 clusters (Fig. 3.3) [42]. This fluorescent property is useful for biosensing and optoelectronics based applications.

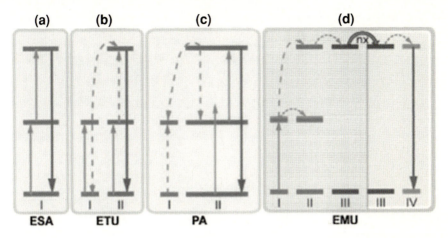

Fig. 3.2 Simplified energy level diagrams describing reported UC processes. ESA, ETU, PA, and EMU denote the excited state absorption, energy transfer UC, photon avalanche, and energy migration-mediated UC, respectively. Reproduced with permission from Ref. [40]. Copyright 2013, Royal Society of Chemistry

Single-walled carbon nanotubes (SWNTs) exhibit intrinsic PL in the NIR light region, known as the desirable "biological transparent window" (700–1,300 nm) where absorption and autofluorescence by tissues, blood, and water are minimized [44–46]. Dai et al. performed ex vivo imaging of tumor slices for 3D reconstruction of the tumor based on the intrinsic fluorescence of SWNTs in the second NIR region (1.1–1.4 μm), revealing the distribution of SWNTs inside the tumor [47].

Carbon nanodots (CNDs) have becoming a promising alternative to traditional luminescent materials in biological and chemical sensing due to their advantages in small size, biocompatibility, low toxicity, and low-cost [19, 48–51]. The synthesis of the CNDs can be generally classified into two kinds of approaches: (1) obtained from a larger natural carbon structure, such as graphite, CNTs; (2) prepared from organic molecular precursor generated carbon materials such as gas soot, candle soot, or activated carbon [48, 52]. CNDs contain many functional groups (such as carboxylic and hydroxyl groups) on the surface, which impart them excellent water solubility and suitability for subsequent functionalization. The emission intensities as well as the color of CNDs can be influenced by the synthesis methods, different excitation wavelengths, particle size, surface properties and other external factors (such as ionic strength and pH values) [48, 52, 53].

3.2 Typical Fluorescence Sensing Strategies

In general, a fluorescence nanosensor system feature has two functional components, i.e., a recognition element to provide specific binding with the targets and a fluorescent nano-transducer component for signaling the binding event. These two

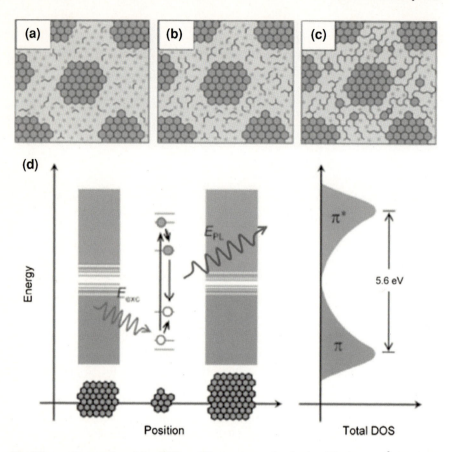

Fig. 3.3 **a–c** Structural models of GO at different stages of reduction. The larger sp^2 clusters of aromatic rings are not drawn to scale. The smaller sp^2 domains indicated by zigzag lines do not necessarily correspond to any specific structure (such as olefinic chains for example) but to small and localized sp^2 configurations that act as the luminescence centers. The PL intensity is relatively weak for (**a**) as-synthesized GO but increases with reduction due to (**b**) formation of additional small sp^2 domains between the larger clusters due to evolution of oxygen with reduction. After extensive reduction, the smaller sp^2 domains create (**c**) percolating pathways among the larger clusters. **d** Representative band structure of GO. The energy levels are quantized with large energy gap for small fragments due to confinement. A photogenerated e–h pair recombining radiatively is depicted. Reproduced with permission from Ref. [42]. Copyright 2010, Wiley

components determine the performance of the nanosensor in terms of selectivity, sensitivity, response time, and signal-to-noise ratio. Based on the fluorescent properties of nanomaterials and requirement of the assay, fluorescent nanosensor-based detection can be typically designed as target induced signal variation, aggregation or anti-aggregation induced signal variation, fluorescence resonance energy transfer (FRET) system, and fluorescence imaging strategy.

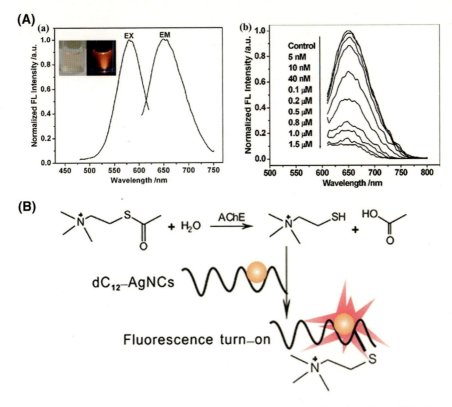

Fig. 3.4 **A** (*a*) The maximum excitation and emission spectra of oligonucleotide-stabilized Ag nanoclusters. *Inset* pictures of Ag nanocluster solution under room light (*left*) and excitation source (*right*) irradiations. (*b*) Fluorescence spectra representing the quenching effect of Hg^{2+} at different concentrations on the fluorescence emission of the Ag nanoclusters (excitation wavelength, 580 nm) with an incubation time of 2 min in an ammonium acetate buffer solution (40 mM, pH 7.0). Reproduced with permission from Ref. [54]. Copyright 2009, Royal Society of Chemistry. **B** Principal reactions of fluorescence assay for AChE activity. Reproduced with permission from Ref. [55]. Copyright 2013, American Chemical Society

3.2.1 Target Induced Signal Variation

The functionalization of fluorescence nanoprobes with ligands that bind target provides a general means to develop turn-on or turn-off fluorescence sensing strategies. The ability of target or target-ligand complexes to quench or enhance the fluorescence nanomaterials provides an intrinsic feature for the operation of these sensors. For example, as shown in Fig. 3.4A, based on the selective quenching effect of Hg^{2+} on Ag nanocluster fluorescence, an oligonucleotide-stabilized Ag nanoclusters fluorescent probe for the determination of Hg^{2+} could be developed [54]. Fluorescence turn-on assay was established using DNA templated silver nanoclusters for rapid, ultrasensitive detection of acetylcholin-esterase (AChE) (Fig. 3.4B) [55]. In this detection, AChE hydrolyzed the acet-ylthiocholine chloride to produce thiocholine (TCh), which could sensitively and rapidly react with nanoclusters via the formation of Ag–S bond and thus enhanced the fluorescence of nanoclusters.

3.2.2 Aggregation or Anti-Aggregation Induced Signal Variation

Nie et al. reported a label-free fluorescent assay for monitoring the activity of protein kinases based on the aggregation behavior of unmodified CdTe QDs (Fig. 3.5a) [56]. Cationic substrate peptides induce the selective aggregation of unmodified QDs with anionic surface charge, whereas phosphorylation by kinase alters the net charge of peptides and then inhibits the aggregation of unmodified QDs, causing an enhanced fluorescence with a 45 nm blue-shift in emission and a yellow-to-green emission color change. The fluorescence response allows this QD-based method to easily probe the activity of cAMP-dependent protein kinase with a low detection limit of 0.47 mU μL^{-1}. Kong et al. developed a novel, selective, and sensitive biosensing system via the reversible dissolution and aggregation of SWNTs directed by the aptamer–protein interaction for the clinical assays of proteins (Fig. 3.5b) [45].

3.2.3 Fluorescence Resonance Energy Transfer System

FRET is a nonradiative process whereby an excited state donor D (usually a fluorophore) transfers energy to a proximal ground state acceptor A through long-range dipole–dipole interactions (Fig. 3.6) [57, 58]. The acceptor must absorb energy at the emission wavelength of the donor but does not necessarily have to remit the energy fluorescently itself. The rate of energy transfer is highly depen-dent on many factors, such as the extent of spectral overlap, the relative orientation of the transition dipoles, and the distance between the donor and acceptor mole-cules [59, 60]. FRET usually occurs over distances comparable to the dimensions of most biological macromolecules (about 1–10 nm).

FRET has been widely used for designing fluorescent nanoprobes. For instance, Chang et al. have demonstrated that the use of two differently sized Au nano-particles acting separately as donor and acceptor, in homogeneous photolumi-nescence quenching assays developed for the analysis of proteins (Fig. 3.7) [61]. In this work, a breast cancer marker protein, platelet-derived growth factor AA (PDGF AA), modified on 11-mercaptoundecanoic acid-protected 2.0 nm photo-luminescent Au nanodots (LAuND) (PDGF AA-LAuND) was employed as the

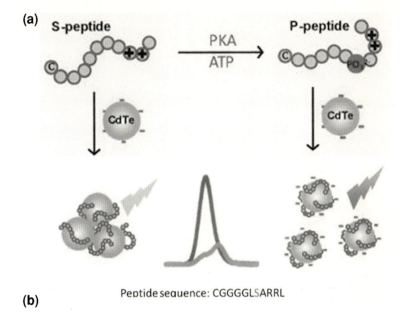

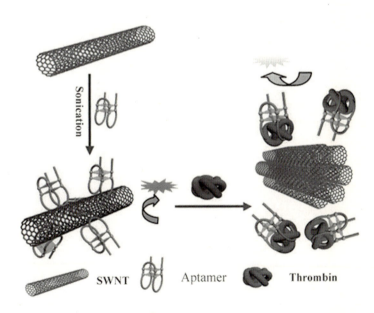

Fig. 3.5 a Concept of the fluorescence kinase activity assay based on QD aggregation. Reproduced with permission from Ref. [56]. Copyright 2010, American Chemical Society. **b** Illustration of SWNT dissolution and aggregation to form a TMB sensor via aptamer–protein interaction. Reproduced with permission from Ref. [45]. Copyright 2009, Royal Society of Chemistry

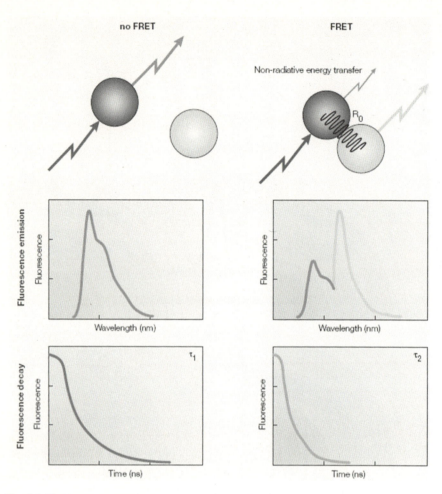

Fig. 3.6 Fluorescence resonance energy transfer (FRET) is the physicochemical phenomenon that is characterized by the transfer of energy from an excited donor chromophore to an acceptor chromophore, without associated radiation release. FRET occurs when the donor emission and acceptor excitation spectra overlap considerably (not shown) and the two dipoles are very close to each other (2–7 nm). FRET is proportional to the sixth power of the distance between the chromophores and, therefore, even minor conformational changes can induce considerable FRET changes. Reproduced with permission from Ref. [58]. Copyright 2003, Nature Publishing Group

donor. Thiol derivative PDGF binding aptamers (Apt) and 13 nm spherical Au NPs (Apt-QAuNP) were used as the acceptor. The photoluminescence of PDGF AA-LAuND at 520 nm decreased due to the photoluminescence of PDGF AA-LAuND quenched by Apt-QAuNP. Based on the PDGF AA-LAuND/Apt-QAuNP molecular light switching system, PDGFs and PDGF α-receptor was analyzed in separate homogeneous solutions. In the presence of PDGFs or PDGF α-receptor,

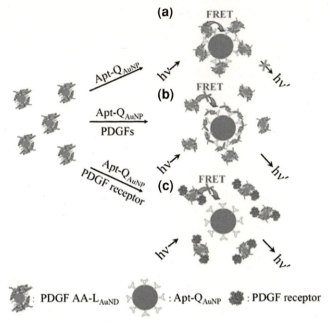

: PDGF AA-L$_{AuND}$: Apt-Q$_{AuNP}$: PDGF receptor

^{a} h, Planck's constant; ν, frequency of light.

Fig. 3.7 Schematic representations of PDGF and PDGF receptor nanosensors. That operate based on the modulation of the photoluminescence quenching between PDGF AA-LAuND and Apt-QAuNP. Reproduced with permission from Ref. [61]. Copyright 2008, American Chemical Society

the interaction between Apt-QAuNP and PDGF AA-LAuND decreased as a result of competitive reactions between the PDGFs and Apt-QAuNP or PDGF α-receptor and PDGF AA-LAuND. The detection limits for PDGF AA and PDGF α-receptor were 80 pM and 0.25 nM, respectively.

3.2.4 Fluorescence Imaging

The use of fluorescence nanoprobes for in vivo imaging has also steadily increased in the last decade. Nie et al. produced QDs-based multifunctional nanoparticle probes for cancer targeting and imaging in live animals [62]. They first encapsulated luminescent QDs with an ABC triblock copolymer and linked this amphiphilic polymer to tumor-targeting ligands. Then they applied this probe to achieve both passive tumor imaging and active tumor imaging with high sensitivity and multicolour capabilities. Key features for the success include both the in vivo imaging sensitivity of QDs and the specific targeting, enhanced permeability and retention of nanoprobes at tumor sites (Fig. 3.8).

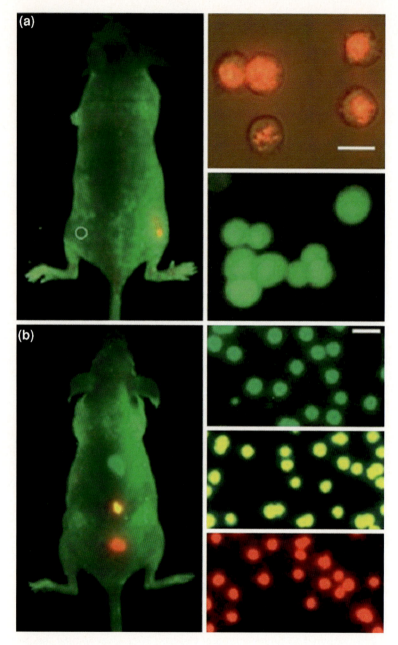

Fig. 3.8 Multi-color QD imaging in live mice. QD-tagged cancer cells **a** *orange, upper* and *green* fluorescent protein (GFP)-labeled cells **a** *green, lower* were injected on the *right flank* and *left flank* (*circle*) of a mouse, respectively. The sensitivity of the QD-tagged cancer cells was comparable to GFP transfected cancer cells. In order to show the *multi-color* imaging ability, QD-encoded microbeads were injected into a mouse **b**. The *right-hand images* showed QD-encoded microbeads emitting *green, yellow* or *red* light. Reproduced with permission from Ref. [62]. Copyright 2004, Nature Publishing Group

3.3 Applications

The attractive fluorescent properties and interesting sensing strategies described above have led to strong interest in the use of nanoprobes for a variety of biological and chemical fluorescence sensing applications.

3.3.1 Heavy Metal Ion Detection

Heavy metal ions are highly toxic species and not biodegradable, therefore they can cause long-term damage to human health and the ecological environment. The development of ultrasensitive fluorescent assays for the detection of heavy metal ions has attracted considerable research efforts in recent years (Table 3.1). Wang et al. reported an efficient ratiometric fluorescence probe based on dual-emission

Table 3.1 Use of nanomaterials for fluorescent sensing of heavy metal ions

Fluorophores	Quencher	Analyte	Matrix	Detection limit	Detection mode	References
Dye	AuNPs	Hg^{2+}	Sodium tetraborate	2.0 ppb	Turn-on	[66]
AuNCs		Hg^{2+} and CH_3Hg^+	Phosphate	Hg^{2+}: 3 pM CH_3Hg^+: 4 nM	Turn-off	[67]
AuNCs	Target	Hg^{2+}	Aqueous	0.5 nM	Turn-off	[68]
AgNCs		Cu^{2+}	Phosphate	8 nM	Turn-on	[69]
Dye	GO	Ag(I)	MOPS buffer	20 nM	Turn-on	[70]
AgNCs	Target	Hg^{2+}	Water	10^{-10} M	Turn-off	[1]
AgNCs	Target	Hg^{2+}	Ammonium acetate	5 nM	Turn-off	[71]
Dye	GO	Pb^{2+}	HEPES	300 pM	Turn-on	[65]
AgNCs	Target	Hg^{2+}		10 nM	Turn-off	[15]
QDs	AuNPs	Hg^{2+}	PBS	0.4–1.2 ppb	Turn-off	[72]
CdSe–ZnS QDs	Target	Hg^{2+}	Water	0.2 ppm	Turn-off	[73]
QDs	Target	Cu^{2+}	HEPES	1.1 nM	Turn-off	[63]
Graphene QDs	Target	Cu^{2+}	Tris–HNO3	6.9 nM	Turn-off	[74]
AgNCs	Target	Cu^{2+}	Ac buffer	10 nM	Turn-off	[75]
Silver–gold alloy NCs		Al^{3+}	Acetate buffer	0.8 μM	Turn-on	[76]
AgNCs	CNTs	Hg^{2+}	PBS	33 pM	Turn-on	[77]
Upconversion NPs		Hg^{2+}	CH_3CN HEPES	8.2 ppb	Turn-on	[64]

NCs nanoclusters; *QDs* quantum dots; *NPs* nanoparticles; *CNTs* carbon nanotubes

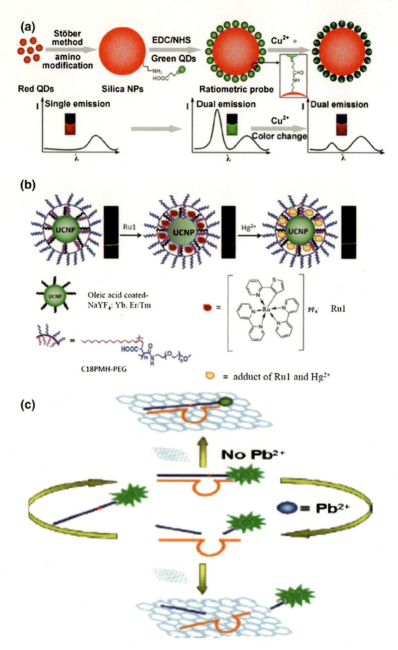

Fig. 3.9 **a** Schematic illustration of the ratiometric probe structure and the visual detection principle for copper ions. Reproduced with permission from Ref. [63]. Copyright 2013, American Chemical Society. **b** Schematic illustration of the synthesis of a UCNP-based nanosystem and its application in detection of Hg^{2+}. Reproduced with permission from Ref. [64]. Copyright 2014, Royal Society of Chemistry. **c** Schematic illustration of the DNAzyme-GO-based fluorescence sensor for Pb^{2+}. Reproduced with permission from Ref. [65]. Copyright 2011, American Chemical Society

QDs hybrid for on-site determination of copper ions with a detection limit of 1.1 nM [63] (Fig. 3.9a). Feng et al. developed cyclometallated ruthenium complex-modified upconversion nanophosphors for selective detection of Hg^{2+} with a detection limit of 8.2 ppb [64] (Fig. 3.9b). Zhang et al. developed a GO-DNA-zyme based biosensor for amplified fluorescence detection of Pb^{2+} [65] (Fig. 3.9c). In this work, the fluorescence of dye was quenched by GO through FRET. A dramatic increase in the fluorescence intensity was observed with the increasing concentrations of Pb^{2+}. Taking advantage of the super fluorescence quenching efficiency of GO, this sensor exhibits high sensitivity with a detection limit of 300 pM.

3.3.2 Small Molecule Detection

The design and construction of fluorescence sensors for recognizing biologically important small molecules have received considerable attention. Many nanomaterials-based fluorescent approaches have been developed, as summarized in Table 3.2.

3.3.3 Nucleic Acid Detection

The fluorescent nanoprobes for nucleic acids detection is mainly based on two well-established scientific foundations: (1) precise DNA hybridization and (2) microenvironment dependent fluorescence properties of nanoprobes. When encountering targets, the change of the DNA configuration in nanoprobes will cause variations in fluorescent signals. Take GO as example, GO can selectively bind with single-strand DNA (ssDNA). When dye labeled ssDNA probes which already bound to the GO surface are hybridized with its complementary target ssDNA, the dye labeled ssDNA probes would detach from GO by forming a DNA duplex, which result in the recovery of fluorescence previously quenched by GO. Based on this concept, Chen et al. reported a graphene platform for sensing aptamer [85] (Fig. 3.10a). Ju et al. developed a simple, highly sensitive, and selective multiple microRNA detection method based on the GO fluorescence quenching and isothermal strand-displacement polymerase reaction with a detection limit of 2.1 fM [86] (Fig. 3.10b). Other types of nucleic acids fluorescence nanosensors have been summarized in Table 3.3.

Table 3.2 Use of nanomaterials for fluorescent sensing of small molecules

Fluorophores	Quencher	Analyte	Matrix	Detection limit	Detection mode	References
Copper NPs	Reporter DNA	ATP	MOPS	28 nM	Turn-off	[78]
Dye	CNTs	ATP	Tris-HCl	4.5 nM	Turn-on	[79]
Fluorescent silica NPs		AMP	SSC1 buffer	0.1 μM	Turn-on	[80]
Mn-doped ZnSe QDs	Target	5-fluorouracil	PBS	128 nM	Turn-off	[81]
Gold NCs	Target	Hydrogen peroxide	glycine buffer	30 nM	Turn-off	[82]
GO	Target	Dopamine	Tris-HCl	94 nM	Turn-off	[43]
Reduced GO decorated with carbon dots	Reactive oxygen species	Acetylcholine	Tris-HCl	30 pM	Turn-off	[83]
Carbon QDs	Target	NO$_2$		250 ppb	Turn-off	[84]

CNTs Carbon nanotubes; *NPs* nanoparticles; *NCs* nanoclusters; *GO* graphene oxide; *QDs* quantum dots

3.3.4 Protein Detection

Many disease states are often associated with the presence of certain biomarker proteins or irregular protein concentrations. Nowadays, several novel nanomaterial-based fluorescence sensing systems have been successfully applied for ƒ detection of proteins (Table 3.4). Min et al. reported a fluorescence method for detection of endonuclease/methyltransferase activity based on the quenching capacity of GO (Fig. 3.11a) [91]. Pang et al. constructed an aptamer biosensor for thrombin detection based on FRET from upconverting phosphors to carbon nanoparticles with a detection limit of 0.18 nM (Fig. 3.11b) [92].

3.3.5 Others

Numerous efforts have been devoted to the development of sensor system for anionic species. Anions detection is challenged by their lower charge to radius ratio, pH sensitivity, wide range of geometries, and solvent dependent binding affinity and selectivity. Chang et al. developed a one-pot approach to prepare fluorescent DNA-templated god/silver nanoclusters (DNA-Au/Ag NCs) for the detection of sulfide ions with a LOD of 0.83 nM [33] (Fig. 3.12).

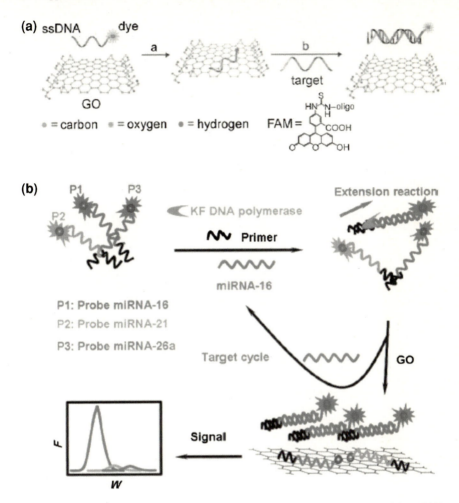

Fig. 3.10 **a** Schematic representation of the target-induced fluorescence change of the ssDNA–FAM–GO complex. Reproduced with permission from Ref. [85]. Copyright 2009, Wiley. **b** Illustration of the GO fluorescence quenching and ISDPR-based multiple miRNA analysis. Reproduced with permission from Ref. [86]. Copyright 2012, American Chemical Society

Selective and sensitive detection of pathogens is imperative as harmful bacteria are the most common causes of food- and waterborne illnesses. Nanomaterial based fluorescence sensors have also been employed to detect pathogens. Chang et al. reported a novel, simple and convenient method by synthesis of fluorescent carbohydrate-protected Au nanodots for the detection of concanavalin A (Con A) and *Escherichia coli* [101] (Fig. 3.13).

Table 3.3 Use of nanomaterials for fluorescent sensing of nucleic acids

Fluorophores	Quencher	Analyte	Matrix	Detection limit	Detection mode	References
Dye	GO	Double-stranded DNA hybridization	Tris-HCl buffer		Turn-on	[85]
Dye	Gold NPs	Single-nucleotide polymorphisms	Tris-HCl buffer		Turn-on	[87]
Dye	Single walled CNTs	DNA hybridization	Tris-HCl buffer	2.3 nM	Turn-on	[88]
Dye	Gold NPs	Sequence-specific DNA	PBS	500 pm	Turn-on	[23]
Silver NCs	MicroRNAs	MicroRNAs			Turn-off	[27]
Dye	GO	Multiple microRNA	Tris-HCl buffer	2.1 fM	Turn-on	[86]
Upconversion NPs	Poly-mphenylenediamine (PMPD) nanospheres	DNA	HEPES buffer	0.036 nM	Turn-on	[89]
Nucleic-acid-stabilized silver NCs	GO	Aptamer and pathogenic DNAs	Phosphate buffer	0.5 nM	Turn-on	[90]

CNTs Carbon nanotubes; *NPs* nanoparticles; *NCs* nanoclusters; *GO* graphene oxide

Table 3.4 Use of nanomaterials for fluorescent sensing of protein

Fluorophores	Quencher	Analyte	Matrix	Detection limit	Detection mode	References
QDs	Oligonuculeotide	Thrombin	Tris-HCl buffer		Turn-on	[20]
Dye	GO	Helicase	Tris-HCl buffer		Turn-off	[93]
PbS QDs	Protein (target)	Thrombin	TAE buffer	1.5 nM	Turn-off	[94]
Single-walled CNTs	Aggregation	Thrombin	Tris-HCl buffer	0.1 nM	Turn-on	[45]
CdTe QDs	Aggregation	cAMP-dependent protein kinase (PKA)	Tris-HCl buffer	$0.47 \text{ mU } \mu L^{-1}$	Turn-on	[56]
Dye	GO	Thrombin	PBS	31.3 pM	Turn-on	[2]
Dye	Gold NPs	Prostate-specific antigen (PSA)		0.032 pg/mL	Turn-on	[95]
Eu^{3+} complex	CNTs	lysozyme	Tris-HCl buffer	0.9 nM	Turn-on	[96]
Pyrene	GO	Anti-HIV-1 gp120 antibody Anti-HIV-2 gp36 antibody	PBS		Turn-on	[97]
Upconversion phosphors	Carbon NPs	Matrix metalloproteinase-2 (MMP-2)	TCNB buffer	10 pg/mL	Turn-on	[98]
Silver nanoclusters		Human α-thrombin	Tris-HCl	1 nM	Turn-on	[99]
Polystyrene nanoparticles loaded with Eu^{3+} chelates (EuNPs)	QDs	Bovine serum albumin	citrate buffer	below 10 ng	Turn-on	[100]
Silver nanoclusters		Acetylcholinesterase activity	phosphate buffer	0.5×10^{-4} U/mL	Turn-on	[55]
Mn-doped ZnS quantum dots	Carbon nanodots	Thrombin	Tris-HCl	0.013 nm	Turn-on	[94]

CNTs: Carbon nanotubes; NPs: na noparticles; NCs: nanoclusters; GO: graphene oxide; QDs: quantum dots

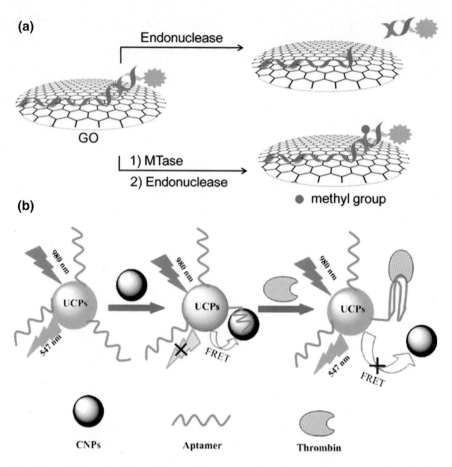

Fig. 3.11 **a** Strategy for ENase/MTase activity assays, based on fluorescence quenching by GO. Reproduced with permission from Ref. [91] Copyright 2011, American Chemical Society. **b** Schematic illustration of the thrombin sensor based on fluorescence resonance energy transfer from aptamer-modified upconverting phosphors to carbon nanoparticles. Reproduced with permission from Ref. [92]. Copyright 2011, American Chemical Society

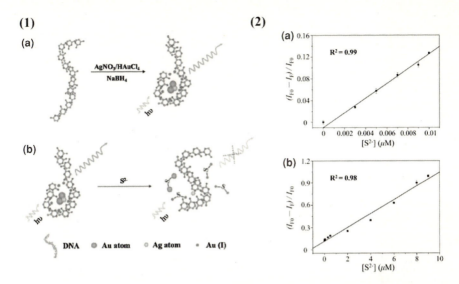

Fig. 3.12 *1* Schematic representation of the preparation and the operation of the DNA-Au/Ag NCs probe for the Detection of S^{2-} Ions: (a) One-pot synthesis of fluorescent DNA-Au/Ag NCs and (b) S^{2-} ions induced fluorescence quenching of the DNA-Au/Ag NCs. *2* Relative fluorescence intensity $[(I_{F0} - I_F)/I_{F0}]$ of DNA-Au/Ag NCs in the presence of S^{2-} ions (0–9 μM) and NaS_2O_8 (0.5 mM). Plots of the values of $[(I_{F0} - I_F)/I_{F0}]$ at 630 nm versus the concentrations of S^{2-} ions over (a) 0–0.01 μM and (b) 0.01–9 μM. Reproduced with permission from Ref. [33]. Copyright 2011, American Chemical Society

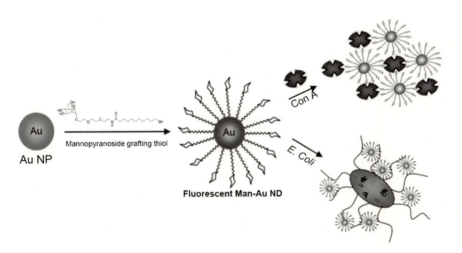

Fig. 3.13 Schematic representation of the preparation of fluorescent Man–Au NDs for the detection of Con A and *E. coli*. Reproduced with permission from Ref. [101]. Copyright 2009, American Chemical Society

References

1. Adhikari B, Banerjee A (2010) Facile synthesis of water-soluble fluorescent silver nanoclusters and HgII sensing. ChemMater 22(15):4364–4371
2. Chang H, Tang L, Wang Y, Jiang J, Li J (2010) Graphene fluorescence resonance energy transfer aptasensor for the thrombin detection. Anal Chem 82(6):2341–2346
3. Chen JL, Yan XP (2011) Ionic strength and pH reversible response of visible and near-infrared fluorescence of graphene oxide nanosheets for monitoring the extracellular pH. Chem Commun 47(11):3135–3137
4. Chi CW, Lao YH, Li YS, Chen LC (2011) A quantum dot-aptamer beacon using a DNA intercalating dye as the FRET reporter: application to label-free thrombin detection. Biosens Bioelectron 26(7):3346–3352
5. Chou CC, Huang YH (2012) Nucleic acid sandwich hybridization assay with quantum dot-induced fluorescence resonance energy transfer for pathogen detection. Sensors 12(12):16660–16672
6. Dong Y, Li G, Zhou N, Wang R, Chi Y, Chen G (2012) Graphene quantum dot as a green and facile sensor for free chlorine in drinking water. Anal Chem 84(19):8378–8382
7. Zhou DM, Xi Q, Liang MF, Chen CH, Tang LJ, Jiang JH (2013) Graphene oxide-hairpin probe nanocomposite as a homogeneous assay platform for DNA base excision repair screening. Biosens Bioelectron 41:359–365
8. Chen Y, Star A, Vidal S (2013) Sweet carbon nanostructures: carbohydrate conjugates with carbon nanotubes and graphene and their applications. Chem Soc Rev 42(11):4532–4542
9. Doane TL, Burda C (2012) The unique role of nanoparticles in nanomedicine: imaging, drug delivery and therapy. Chem Soc Rev 41(7):2885–2911
10. Kershaw SV, Susha AS, Rogach AL (2013) Narrow bandgap colloidal metal chalcogenide quantum dots: synthetic methods, heterostructures, assemblies, electronic and infrared optical properties. Chem Soc Rev 42(7):3033–3087
11. Wu P, Yan XP (2013) Doped quantum dots for chemo/biosensing and bioimaging. Chem Soc Rev 42(12):5489–5521
12. Yong KT, Law WC, Hu R, Ye L, Liu L, Swihart MT, Prasad PN (2013) Nanotoxicity assessment of quantum dots: from cellular to primate studies. Chem Soc Rev 42(3):1236–1250
13. Zhou J, Liu Z, Li F (2012) Upconversion nanophosphors for small-animal imaging. Chem Soc Rev 41(3):1323–1349
14. Dabbousi BO, Rodriguez-Viejo J, Mikulec FV, Heine JR, Mattoussi H, Ober R, Jensen KF, Bawendi MG (1997) (CdSe)ZnS core–shell quantum dots: synthesis and characterization of a size series of highly luminescent nanocrystallites. J Phys Chem B 101(46):9463–9475
15. Guo C, Irudayaraj J (2011) Fluorescent Ag clusters via a protein-directed approach as a Hg(II) ion sensor. Anal Chem 83(8):2883–2889
16. Xie J, Zheng Y, Ying JY (2009) Protein-directed synthesis of highly fluorescent gold nanoclusters. J Am Chem Soc 131(3):888–889
17. Wang F, Liu X (2008) Upconversion multicolor fine-tuning: visible to near-infrared emission from lanthanide-doped NaYF$_4$ nanoparticles. J Am Chem Soc 130(17):5642–5643
18. Adeli M, Soleyman R, Beiranvand Z, Madani F (2013) Carbon nanotubes in cancer therapy: a more precise look at the role of carbon nanotube–polymer interactions. Chem Soc Rev 42(12):5231–5256
19. Liu H, Ye T, Mao C (2007) Fluorescent carbon nanoparticles derived from candle soot. Angew Chem Int Ed 46(34):6473–6475
20. Levy M, Cater SF, Ellington AD (2005) Quantum-dot aptamer beacons for the detection of proteins. ChemBioChem 6(12):2163–2166
21. Li M, Wang Q, Shi X, Hornak LA, Wu N (2011) Detection of mercury(II) by quantum dot/DNA/gold nanoparticle ensemble based nanosensor via nanometal surface energy transfer. Anal Chem 83(18):7061–7065

22. Li T, Zhou Y, Sun J, Tang D, Guo S, Ding X (2011) Ultrasensitive detection of mercury(II) ion using CdTe quantum dots in sol-gel-derived silica spheres coated with calix[6]arene as fluorescent probes. Microchim Acta 175(1–2):113–119
23. Song S, Liang Z, Zhang J, Wang L, Li G, Fan C (2009) Gold-nanoparticle-based multicolor nanobeacons for sequence-specific DNA analysis. Angew Chem Int Ed 48(46):8670–8674
24. Han E, Ding L, Ju H (2011) Highly sensitive fluorescent analysis of dynamic glycan expression on living cells using glyconanoparticles and functionalized quantum dots. Anal Chem 83(18):7006–7012
25. Huang Y, Zhao S, Shi M, Chen J, Chen ZF, Liang H (2011) Intermolecular and intramolecular quencher based quantum dot nanoprobes for multiplexed detection of endonuclease activity and inhibition. Anal Chem 83(23):8913–8918
26. Xia Y, Song L, Zhu C (2011) Turn-on and near-infrared fluorescent sensing for 2,4,6-trinitrotoluene based on hybrid (gold nanorod)-(quantum dots) assembly. Anal Chem 83(4):1401–1407
27. Yang SW, Vosch T (2011) Rapid detection of microRNA by a silver nanocluster DNA probe. Anal Chem 83(18):6935–6939
28. Shang L, Dong S, Nienhaus GU (2011) Ultra-small fluorescent metal nanoclusters: synthesis and biological applications. Nano Today 6(4):401–418
29. Kawasaki H, Hamaguchi K, Osaka I, Arakawa R (2011) pH-dependent synthesis of pepsin-mediated gold nanoclusters with blue green and red fluorescent emission. Adv Funct Mater 21(18):3508–3515
30. Yang X, Shi M, Zhou R, Chen X, Chen H (2011) Blending of HAuCl$_4$ and histidine in aqueous solution: a simple approach to the Au$_{10}$ cluster. Nanoscale 3(6):2596–2601
31. Diez I, Ras RH (2011) Fluorescent silver nanoclusters. Nanoscale 3(5):1963–1970
32. Polavarapu L, Manna M, Xu QH (2011) Biocompatible glutathione capped gold clusters as one- and two-photon excitation fluorescence contrast agents for live cells imaging. Nanoscale 3(2):429–434
33. Chen WY, Lan GY, Chang HT (2011) Use of fluorescent DNA-templated gold/silver nanoclusters for the detection of sulfide ions. Anal Chem 83(24):9450–9455
34. Liu Y, Ai K, Cheng X, Huo L, Lu L (2010) Gold-nanocluster-based fluorescent sensors for highly sensitive and selective detection of cyanide in water. Adv Funct Mater 20(6):951–956
35. Muhammed MA, Verma PK, Pal SK, Kumar RC, Paul S, Omkumar RV, Pradeep T (2009) Bright, NIR-emitting Au$_{23}$ from Au$_{25}$: characterization and applications including biolabeling. Chemistry 15(39):10110–10120
36. Duan H, Nie S (2007) Etching colloidal gold nanocrystals with hyperbranched and multivalent polymers: a new route to fluorescent and water-soluble atomic clusters. J Am Chem Soc 129(9):2412–2413
37. Lin H, Li L, Lei C, Xu X, Nie Z, Guo M, Huang Y, Yao S (2013) Immune-independent and label-free fluorescent assay for cystatin c detection based on protein-stabilized Au nanoclusters. Biosens Bioelectron 41:256–261
38. Li Z, Zhang Y (2008) An efficient and user-friendly method for the synthesis of hexagonal-phase NaYF$_4$:Yb, Er/Tm nanocrystals with controllable shape and upconversion fluorescence. Nanotechnology 19(34):345606
39. Mai H-X, Zhang Y-W, Si R, Yan Z-G, Sun L-d, You L-P, Yan C-H (2006) High-quality sodium rare-earth fluoride nanocrystals: controlled synthesis and optical properties. J Am Chem Soc 128(19):6426–6436
40. Liu Y, Tu D, Zhu H, Chen X (2013) Lanthanide-doped luminescent nanoprobes: controlled synthesis, optical spectroscopy, and bioapplications. Chem Soc Rev 42(16):6924–6958
41. Loh KP, Bao Q, Eda G, Chhowalla M (2010) Graphene oxide as a chemically tunable platform for optical applications. Nat Chem 2(12):1015–1024
42. Eda G, Lin Y-Y, Mattevi C, Yamaguchi H, Chen H-A, Chen IS, Chen C-W, Chhowalla M (2010) Blue photoluminescence from chemically derived graphene oxide. Adv Mater 22(4):505–509

43. Chen JL, Yan XP, Meng K, Wang SF (2011) Graphene oxide based photoinduced charge transfer label-free near-infrared fluorescent biosensor for dopamine. Anal Chem 83(22):8787–8793
44. Barone PW, Strano MS (2006) Reversible control of carbon nanotube aggregation for a glucose affinity sensor. Angew Chem Int Ed 45(48):8138–8141
45. Chen H, Yu C, Jiang C, Zhang S, Liu B, Kong J (2009) A novel near-infrared protein assay based on the dissolution and aggregation of aptamer-wrapped single-walled carbon nanotubes. Chem Commun 33:5006–5008
46. O'Connell MJ, Bachilo SM, Huffman CB, Moore VC, Strano MS, Haroz EH, Rialon KL, Boul PJ, Noon WH, Kittrell C, Ma J, Hauge RH, Weisman RB, Smalley RE (2002) Band gap fluorescence from individual single-walled carbon nanotubes. Science 297(5581):593–596
47. Robinson JT, Hong G, Liang Y, Zhang B, Yaghi OK, Dai H (2012) In vivo fluorescence imaging in the second near-infrared window with long circulating carbon nanotubes capable of ultrahigh tumor uptake. J Am Chem Soc 134(25):10664–10669
48. Zhang C, Lin J (2012) Defect-related luminescent materials: synthesis, emission properties and applications. Chem Soc Rev 41(23):7938–7961
49. Lin Z, Xue W, Chen H, Lin JM (2011) Peroxynitrous-acid-induced chemiluminescence of fluorescent carbon dots for nitrite sensing. Anal Chem 83(21):8245–8251
50. Qu K, Wang J, Ren J, Qu X (2013) Carbon dots prepared by hydrothermal treatment of dopamine as an effective fluorescent sensing platform for the label-free detection of iron(III) ions and dopamine. Chemistry 19(22):7243–7249
51. Zhang L, Cui P, Zhang B, Gao F (2013) Aptamer-based turn-on detection of thrombin in biological fluids based on efficient phosphorescence energy transfer from Mn-doped ZnS quantum dots to carbon nanodots. Chemistry 19(28):9242–9250
52. Hsu P-C, Shih Z-Y, Lee C-H, Chang H-T (2012) Synthesis and analytical applications of photoluminescent carbon nanodots. Green Chem 14(4):917
53. Li H, Kang Z, Liu Y, Lee S-T (2012) Carbon nanodots: synthesis, properties and applications. J Mater Chem 22(46):24230
54. Guo W, Yuan J, Wang E (2009) Oligonucleotide-stabilized Ag nanoclusters as novel fluorescence probes for the highly selective and sensitive detection of the Hg^{2+} ion. ChemCommun 23:3395–3397
55. Zhang Y, Cai Y, Qi Z, Lu L, Qian Y (2013) DNA-templated silver nanoclusters for fluorescence turn-on assay of acetylcholinesterase activity. Anal Chem 85(17):8455–8461
56. Xu X, Liu X, Nie Z, Pan Y, Guo M, Yao S (2010) Label-free fluorescent detection of protein kinase activity based on the aggregation behavior of unmodified quantum dots. Anal Chem 83(1):52–59
57. Sapsford KE, Berti L, Medintz IL (2006) Materials for fluorescence resonance energy transfer analysis: beyond traditional donor-acceptor combinations. Angew Chem Int Ed 45(28):4562–4589
58. Rudolf R, Mongillo M, Rizzuto R, Pozzan T (2003) Looking forward to seeing calcium. Nat Rev Mol Cell Biol 4(7):579–586
59. Jares-Erijman EA, Jovin TM (2003) FRET imaging. Nat Biotechnol 21(11):1387–1395
60. Clapp AR, Medintz IL, Fisher BR, Anderson GP, Mattoussi H (2005) Can luminescent quantum dots be efficient energy acceptors with organic dye donors? J Am Chem Soc 127(4):1242–1250
61. Huang C-C, Chiang C-K, Lin Z-H, Lee K-H, Chang H-T (2008) Bioconjugated gold nanodots and nanoparticles for protein assays based on photoluminescence quenching. Anal Chem 80(5):1497–1504
62. Gao X, Cui Y, Levenson RM, Chung LW, Nie S (2004) In vivo cancer targeting and imaging with semiconductor quantum dots. Nat Biotechnol 22(8):969–976
63. Yao J, Zhang K, Zhu H, Ma F, Sun M, Yu H, Sun J, Wang S (2013) Efficient ratiometric fluorescence probe based on dual-emission quantum dots hybrid for on-site determination of copper ions. Anal Chem 85(13):6461–6468

64. Li X, Wu Y, Liu Y, Zou X, Yao L, Li F, Feng W (2014) Cyclometallated ruthenium complex-modified upconversion nanophosphors for selective detection of Hg^{2+} ions in water. Nanoscale 6(2):1020–1028

65. Zhao XH, Kong RM, Zhang XB, Meng HM, Liu WN, Tan W, Shen GL, Yu RQ (2011) Graphene-DNAzyme based biosensor for amplified fluorescence "turn-on" detection of Pb^{2+} with a high selectivity. Anal Chem 83(13):5062–5066

66. Huang C-C, Chang H-T (2006) Selective gold-nanoparticle-based "turn-on" fluorescent sensors for detection of mercury(II) in aqueous solution. Anal Chem 78(24):8332–8338

67. Lin Y-H, Tseng W-L (2010) Ultrasensitive sensing of Hg^{2+} and CH3Hg$^+$ based on the fluorescence quenching of lysozyme type VI-stabilized gold nanoclusters. Anal Chem 82(22):9194–9200

68. Xie J, Zheng Y, Ying JY (2010) Highly selective and ultrasensitive detection of Hg^{2+} based on fluorescence quenching of Au nanoclusters by Hg^{2+}–Au$^+$ interactions. Chem Commun 46(6):961–963

69. Lan GY, Huang CC, Chang HT (2010) Silver nanoclusters as fluorescent probes for selective and sensitive detection of copper ions. Chem Commun 46(8):1257–1259

70. Wen Y, Xing F, He S, Song S, Wang L, Long Y, Li D, Fan C (2010) A graphene-based fluorescent nanoprobe for silver(I) ions detection by using graphene oxide and a silver-specific oligonucleotide. Chem Commun 46(15):2596–2598

71. Guo W, Yuan J, Wang E (2009) Oligonucleotide-stabilized Ag nanoclusters as novel fluorescence probes for the highly selective and sensitive detection of the Hg^{2+} ion. Chem Commun 23:3395–3397

72. Li M, Wang Q, Shi X, Hornak LA, Wu N (2011) Detection of mercury(II) by quantum dot/DNA/gold nanoparticle ensemble based nanosensor via nanometal surface energy transfer. Anal Chem 83(18):7061–7065

73. Yuan C, Zhang K, Zhang Z, Wang S (2012) Highly selective and sensitive detection of mercuric ion based on a visual fluorescence method. Anal Chem 84(22):9792–9801

74. Sun H, Gao N, Wu L, Ren J, Wei W, Qu X (2013) Highly photoluminescent amino-functionalized graphene quantum dots used for sensing copper ions. Chem Eur J 19(40):13362–13368

75. Yuan Z, Cai N, Du Y, He Y, Yeung ES (2014) Sensitive and selective detection of copper ions with highly stable polyethyleneimine-protected silver nanoclusters. Anal Chem 86(1):419–426

76. Zhou TY, Lin LP, Rong MC, Jiang YQ, Chen X (2013) Silver–gold alloy nanoclusters as a fluorescence-enhanced probe for aluminum ion sensing. Anal Chem 85(20):9839–9844

77. Wang G, Xu G, Zhu Y, Zhang X (2014) A "turn-on" carbon nanotube-Ag nanoclusters fluorescent sensor for sensitive and selective detection of Hg^{2+} with cyclic amplification of exonuclease III activity. Chem Commun 50(6):747–750

78. Zhou Z, Du Y, Dong S (2011) Double-strand DNA-templated formation of copper nanoparticles as fluorescent probe for label-free aptamer sensor. Anal Chem 83(13):5122–5127

79. Zhang L, Wei H, Li J, Li T, Li D, Li Y, Wang E (2010) A carbon nanotubes based ATP apta-sensing platform and its application in cellular assay. Biosens Bioelectron 25(8):1897–1901

80. Song Y, Zhao C, Ren J, Qu X (2009) Rapid and ultra-sensitive detection of AMP using a fluorescent and magnetic nano-silica sandwich complex. Chem Commun 15:1975–1977

81. Zhu D, Chen Y, Jiang L, Geng J, Zhang J, Zhu JJ (2011) Manganese-doped ZnSe quantum dots as a probe for time-resolved fluorescence detection of 5-fluorouracil. Anal Chem 83(23):9076–9081

82. Wen F, Dong Y, Feng L, Wang S, Zhang S, Zhang X (2011) Horseradish peroxidase functionalized fluorescent gold nanoclusters for hydrogen peroxide sensing. Anal Chem 83(4):1193–1196

83. Wang CI, Periasamy AP, Chang HT (2013) Photoluminescent C-dots@RGO probe for sensitive and selective detection of acetylcholine. Anal Chem 85(6):3263–3270

84. Wang R, Li G, Dong Y, Chi Y, Chen G (2013) Carbon quantum dot-functionalized aerogels for NO$_2$ gas sensing. Anal Chem 85(17):8065–8069

85. Lu CH, Yang HH, Zhu CL, Chen X, Chen GN (2009) A graphene platform for sensing biomolecules. Angew Chem Int Ed 48(26):4785–4787

86. Dong H, Zhang J, Ju H, Lu H, Wang S, Jin S, Hao K, Du H, Zhang X (2012) Highly sensitive multiple microRNA detection based on fluorescence quenching of graphene oxide and isothermal strand-displacement polymerase reaction. Anal Chem 84(10):4587–4593

87. Wang H, Li J, Wang Y, Jin J, Yang R, Wang K, Tan W (2010) Combination of DNA ligase reaction and gold nanoparticle-quenched fluorescent oligonucleotides: a simple and efficient approach for fluorescent assaying of single-nucleotide polymorphisms. Anal Chem 82(18):7684–7690

88. Liu Y, Wang Y, Jin J, Wang H, Yang R, Tan W (2009) Fluorescent assay of DNA hybridization with label-free molecular switch: reducing background-signal and improving specificity by using carbon nanotubes. Chem Commun 6:665-667

89. Wang Y, Wu Z, Liu Z (2013) Upconversion fluorescence resonance energy transfer biosensor with aromatic polymer nanospheres as the lable-free energy acceptor. Anal Chem 85(1):258–264

90. Liu X, Wang F, Aizen R, Yehezkeli O, Willner I (2013) Graphene oxide/nucleic-acid-stabilized silver nanoclusters: functional hybrid materials for optical aptamer sensing and multiplexed analysis of pathogenic DNAs. J Am Chem Soc 135(32):11832–11839

91. Lee J, Kim YK, Min DH (2011) A new assay for endonuclease/methyltransferase activities based on graphene oxide. Anal Chem 83(23):8906–8912

92. Wang Y, Bao L, Liu Z, Pang DW (2011) Aptamer biosensor based on fluorescence resonance energy transfer from upconverting phosphors to carbon nanoparticles for thrombin detection in human plasma. Anal Chem 83(21):8130–8137

93. Jang H, Kim Y-K, Kwon H-M, Yeo W-S, Kim D-E, Min D-H (2010) A graphene-based platform for the assay of duplex-DNA unwinding by helicase. Angew Chem Int Ed 122(33):5839–5843

94. Zhang L, Cui P, Zhang B, Gao F (2013) Aptamer-based turn-on detection of thrombin in biological fluids based on efficient phosphorescence energy transfer from Mn-doped ZnS quantum dots to carbon nanodots. Chemistry 19(28):9242–9250

95. Liu D, Huang X, Wang Z, Jin A, Sun X, Zhu L, Wang F, Ma Y, Niu G, Hight Walker AR, Chen X (2013) Gold nanoparticle-based activatable probe for sensing ultralow levels of prostate-specific antigen. ACS Nano 7(6):5568–5576

96. Ouyang X, Yu R, Jin J, Li J, Yang R, Tan W, Yuan J (2011) New strategy for label-free and time-resolved luminescent assay of protein: conjugate Eu^{3+} complex and aptamer-wrapped carbon nanotubes. Anal Chem 83(3):782–789

97. Lu CH, Li J, Zhang XL, Zheng AX, Yang HH, Chen X, Chen GN (2011) General approach for monitoring peptide-protein interactions based on graphene-peptide complex. Anal Chem 83(19):7276–7282

98. Wang Y, Shen P, Li C, Wang Y, Liu Z (2012) Upconversion fluorescence resonance energy transfer based biosensor for ultrasensitive detection of matrix metalloproteinase-2 in blood. Anal Chem 84(3):1466–1473

99. Li J, Zhong X, Zhang H, Le XC, Zhu JJ (2012) Binding-induced fluorescence turn-on assay using aptamer-functionalized silver nanocluster DNA probes. Anal Chem 84(12):5170–5174

100. Harma H, Pihlasalo S, Cywinski PJ, Mikkonen P, Hammann T, Lohmannsroben HG, Hanninen P (2013) Protein quantification using resonance energy transfer between donor nanoparticles and acceptor quantum dots. Anal Chem 85(5):2921–2926

101. Huang C-C, Chen C-T, Shiang Y-C, Lin Z-H, Chang H-T (2009) Synthesis of fluorescent carbohydrate-protected Au nanodots for detection of concanavalin A and *Escherichia coli*. Anal Chem 81(3):875–882

Chapter 4
Surface-Enhanced Raman Scattering Nanoprobes

Abstract During the last decade, novel nanoprobes named surface-enhanced Raman scattering (SERS) nanoprobes have drawn much attention of chemists. SERS nanoprobes produce strong, characteristic Raman signals, demonstrating optical labeling capability similar to those of commonly reported organic dyes and fluorescent quantum dots. However, this new generation of optical nanoprobes has the ultra-sensitivity, multiplexing, and quantitative abilities of the SERS technique, and shows extraordinary features for bioanalysis. In this chapter, we focus on the most recent advances of SERS nanoprobes. A brief overview of the basic concept and the synthesis of SERS nanoprobes is given. Also we discuss the nanoprobes' growing popularity for bioanalysis at different levels of molecular multiplex detection, for bacterial and live-cell sensing, and for in vivo imaging from the aspect of different sensing mechanisms.

Keywords Surface-enhanced Raman scattering · Optical nanoprobes · Multiplex detection · Cell labeling · In vivo imaging

4.1 Introduction to Surface-Enhanced Raman Scattering Nanoprobes

SERS was discovered on the rough surface of gold electrode in the 1970s [1, 2]. With the progress of nanoscience and the enrichment of nanomaterials, it is found that ultra-sensitive vibrational spectroscopic Raman signals can also be detected with molecules on or near the surface of many plasmonic nanostructures. SERS phenomenon is attributed to the long-range electromagnetic (EM) enhancement and the short-range chemical enhancement (CE), which has been elaborated in many reviews [3–6]. SERS inherited the advantages of in situ, noninvasive detection and vibrational spectroscopic fingerprint information of conventional Raman technique, while greatly increased the sensitivity up to 10^{14} that suitable for single molecule detection under some conditions [7, 8]. SERS has greatly

L. Chen et al., *Novel Optical Nanoprobes for Chemical and Biological Analysis*, SpringerBriefs in Molecular Science, DOI: 10.1007/978-3-662-43624-0_4, © The Author(s) 2014

extended the role of standard Raman spectroscopy in biochemistry and life sciences [9–11]. In the last 40 years, the classic application was the direct sensing SERS spectra of analytes using metallic SERS substrates to obtain both qualitative and quantitative information [12].

During the last decade, SERS has been applied to design novel nanoprobes named "SERS nanoprobes" [13]. Such SERS-active nanoprobes produce strong, characteristic Raman signals, demonstrating optical labeling capability similar to those of commonly reported organic dyes and fluorescent quantum dots (QDs). However, this new generation of probe has the ultra-sensitivity, multiplexing, and quantitative abilities of the SERS technique, and shows extraordinary features for bioanalysis [11].

4.2 Optical Properties of SERS Nanoprobes

We will first give a brief theoretical guideline for the design and synthesis of SERS nanoprobes by introducing an equation proposed by Kneipp based on these theories [4, 7, 14], in which SERS' Stokes signal, $P^{SERS}(v_s)$, can be estimated:

$$P^{SERS}(v_S) = N\sigma_{ads}^R |A(v_L)|^2 |A(v_S)|^2 I(v_L)$$

Here, $I(v_L)$ is the excitation laser intensity; σ_{ads}^R is the Raman cross-section of the adsorbed molecule, possibly increased due to chemical enhancement; N is the number of molecules that undergo the SERS process; and $A(v_L)$ and $A(v_S)$ are laser and Raman scattering field enhancement factors, respectively. From this equation, it can be seen that the signal-enhanced ability of metal substrates, chemical structure of reporter molecules, and the attached molecule numbers will all influence the sensitivity of SERS nanoprobes.

SERS nanoprobe is typically composed of four parts: a metal nanosubstrate, an organic Raman reporter molecule, a protection shell, and targeting molecules (Fig. 4.1). Various metal nanostructures can provide strongly enhanced spectroscopic signals due to the local optical fields at metal surfaces, which provide rigid foundations for the nanoprobes. In general, their size distribution, geometry, chemical composition, and surface chemistry can influence the Raman enhancement ability. Up to now, various kinds of metal NPs such as gold and silver NPs, gold nanorods, gold nanoflowers, and Au–Ag bimetallic NPs have been reported to serve as substrates of SERS nanoprobes [13]. Furthermore, Raman reporter molecules should be attached to the metal nanostructure to generate SERS fingerprint signatures [15]. To prepare nanoprobes with strong, stable multiplex signals, Raman reporters are required to have large Raman cross-section and affiliation to noble metal surface through electrostatic or covalent binding. Typically, some nitrogen-containing cationic dyes, sulfur-containing dyes, and thio-small molecules can be selected as Raman reporters. Third, carefully designed coating

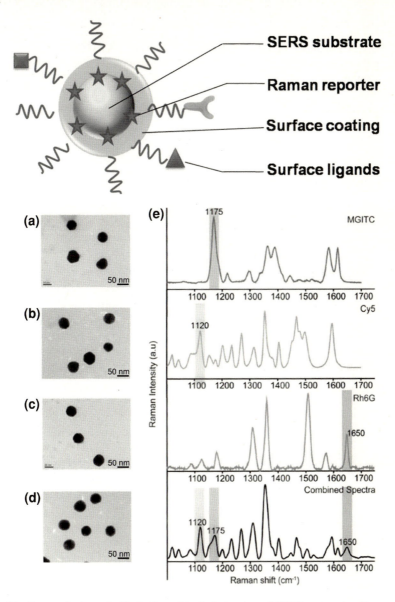

Fig. 4.1 *Up panel* schematic illustration of typical structure of SERS nanoprobe. *Down panel* **a–d** TEM image of the pure AuNPs, SERS nanonanoprobes with MGITC, Cy5 and Rh6G reporter molecules, respectively. **e** SERS spectra of individual SERS nanonanoprobes and its mixture depicting their multiplexing peaks. The most distinctive multiplex peak from each reporter is marked. Reproduced with permission from Ref. [15]. Copyright 2014, Nature Publishing Group

Table 4.1 Comparison of SERS nanoprobes, quantum dots, and conventional dyes

Properties	SERS nanoprobes	Quantum dots	Conventional dyes
Physical principle	Raman scattering	Fluorescence emission	Electronic absorption/ Fluorescence emission
Core composition	Au and Ag based NPs	CdSe and CdTe based NPs	Organic compounds
Size	∼50 nm	∼10 nm	∼1 nm
Bandwidth	<2 nm	∼30–50 nm	Usually >50 nm
Structural information	Fingerprint	Non-fingerprint	Non-fingerprint
Multiplexing capacity	∼10–100	∼3–10	∼1–3
Photostability	Anti-photobleaching	Decay under strong laser	Decay under weak excitation
Toxicity	Not toxic	Toxic	Toxic

Reproduced with permission from Ref. [13]. Copyright 2013, American Chemical Society

materials including biomolecules [16], polymers [17], liposomes [18, 19], and silica shells [20, 21] are essential to improve the nanoprobes' colloidal and signal stability as well as biocompatiblity in biological conditions. Further adding targeting molecules to the SERS nanoprobes imparts biofunctionality.

The development of SERS nanoprobes can be considered as a significant step forward in the spectroscopic analysis. They offer several advantages over fluorescent probes such as organic dyes and QDs (Table 4.1). First, Raman produces vibrational spectral bands with narrow line widths (∼1 nm), [22] thus, Raman-based probes are inherently suitable for multiplex analysis. Second, SERS nanoprobes can provide sufficient sensitivity similar to or better than those generated from fluorescence for trace analysis in some circumstance [23]. Third, the extremely short lifetimes of Raman scattering prevent photobleaching, energy transfer, or quenching of reporters in the excited state, [24] rendering high photostability to SERS nanoprobes. And fourth, optimal contrast can be achieved by using red to near-infrared (NIR) excitation, enabling SERS nanoprobes to be used for noninvasive imaging in living subjects [25].

4.3 Applications

SERS nanoprobes have been widely applied to bio- and chemical analysis in the field of ion and biomolecule detection, cell labeling, tissue diagnosis and in vivo imaging. We will give a detailed introduction of some featured applications from the aspect of different sensing mechanisms including: (1) analyte-induced SERS nanoprobe aggregation/anti-aggregation, (2) SERS-nanoprobe based optical labeling, and (3) analytes induced alteration of the reporter's Raman signature.

Fig. 4.2 a Schematic diagram of the indirect SERS method of measuring melamine using MPY modified AuNPs based on SERS nanoprobe aggregation principle. Reproduced with permission from Ref. [26]. Copyright 2011, Springer. **b** Schematic representation of the anti-aggregation SERS sensing principle for protease detection. Reproduced with permission from Ref. [30]. Copyright 2013, Royal Society of Chemistry

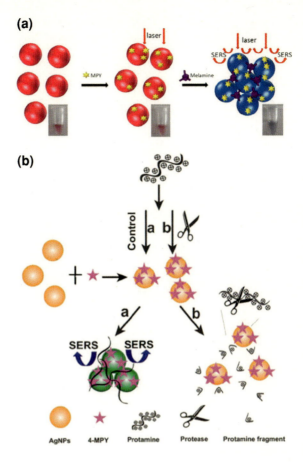

4.3.1 Aggregation or Anti-aggregation Induced Signal Variation

The idea of nanoprobe aggregation based analysis is originated from the greatly enhanced electromagnetic field in the hotspots of noble metal NP junctions. The aggregation can be formed mainly in two mechanisms. The first is that coordinating interactions will neutralize the nanoprobes after addition of certain kinds of analytes with high affinity to metal surfaces. For example, AuNPs aggregated upon the addition of melamine, based on this feature, SERS nanoprobe was developed for rapid and sensitive detection of melamine in milk powder (Fig. 4.2a). The addition amount of melamine can be revealed by significantly enhanced Raman intensity of the reporter molecule 4-mercaptopyridine (4-MPY). The LOD was found to be as low as 0.1 ppb of melamine, with an excellent linearity of 0.5–100 ppb [26].

The second mechanism is to co-modify the Raman reporter and selective ligand on SERS substrates. The targets can induce aggregation specifically and detected in a signal "turn on" mode. Several metal ions have been reported to be detected by this strategy. A highly sensitive SERS platform for the selective trace analysis of As^{3+} ions was reported based on glutathione (GSH)/4-MPY modified AgNPs. GSH conjugated on the surface of AgNPs for specifical binding with As^{3+} ions in aqueous solution through As-O linkage and 4-MPY was used as a Raman reporter. The binding of As^{3+} with GSH resulted in the aggregation of AgNPs, and Raman signal of 4-MPY reporters increased. The LOD could be as low as 0.76 ppb [27]. SERS nanoprobes for the sensitive and selective detection of Cd^{2+} were also reported by taking advannnanoprobee of the interparticle plasmonic coupling generated in the process of Cd^{2+}-selective nanoparticle self-aggregation [28]. Gold NPs were modified with thiolated dyes and a layer of Cd^{2+}-chelating polymer brush coating. Addition of Cd^{2+} leads to interparticle self-aggregation and resulant 90-fold of SERS signal enhancement of reporters. Similarly, Cr^{3+} was detected by using Tween 20 stabilized citrate-capped AuNPs SERS nanoprobes with 2-aminothiophenol as Raman reporter. Due to the chelation between Cr^{3+} and citrate ions, SERS nanoprobes undergo aggregation and the signal of reporters increased regularly. This nanoprobe could recognize Cr^{3+} at a 5×10^{-8} M level in an aqueous medium at a pH of 6.0. The selectivity toward Cr^{3+} was 400-fold remarkably greater than other metal ions [29].

The sensing principle mentioned above is NP aggregation induce SERS signal enhancement. On the contrary, if the analytes can preclude the aggregation process or break down the as-formed NPs clusters into single NP, their addition will decrease SERS intensity. Therefore, an anti-aggregation SERS sensing mode could be established and the targets can be detected in a signal "turn off" way.

A simple and sensitive SERS strategy was developed for recognition and detection of trypsin, by using anti-aggregation of 4-MPY functionalized AgNPs based on the interaction between protamine and trypsin (Fig. 4.2b). In this case, polycationic protamine not only served as a substrate for enzyme hydrolysis but also worked as a medium for SERS enhancement, which could bind negatively charged SERS nanoprobes and induce their aggregation. The hydrolysis catalyzed with trypsin in sample solution decreased the concentration of free protamine, resulting in the dispersion of AgNPs and thus decreasing the Raman intensity of 4-MPY, by which the trypsin could be sensed optically [30]. In a further work, it was found that in the presence of heparin, the interaction between heparin and protamine decreased the concentration of free protamine, which dissipated the aggregated 4-MPY functionalized Ag NPs and thus decreased Raman enhancement effect. The degree of aggregation and Raman enhancement effect was proportional to the concentration of added heparin. Under optimized assay conditions, good linear relationship was obtained over the range of 0.5–150 ng/mL with a minimum detectable concentration of 0.5 ng/mL [31].

The anti-aggregation idea was applied for intracellular drug release investigation tactfully. Song et al. developed bioconjugated SERS active plasmonic vesicles with a hollow cavity assembled from SERS-encoded amphiphilic gold NPs

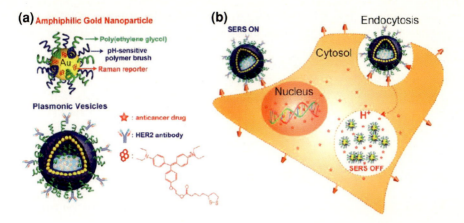

Fig. 4.3 a Schematic illustration of the amphiphilic gold nanoparticle coated with Raman reporter BGLA and mixed polymer brushes of hydrophilic PEG and pH-sensitive hydrophobic PMMAVP grafts and the drug-loaded plasmonic vesicle nanoprobeged with HER2 antibody for cancer cell targeting. **b** The cellular binding, uptake, and intraorganelle disruption of the SERS-encoded pH-sensitive plasmonic vesicles. Reproduced with permission from Ref. [32]. Copyright 2012, American Chemical Society

(Fig. 4.3). The structure of the assemblies was pH-responsive, it disassembled back to the plasmonic vesicle, stimulated by the hydrophobic-to-hydrophilic transition of the hydrophobic brushes in acidic intracellular compartments, allows for triggered intracellular drug release. Disassembly of the vesicles also leads to dramatic decrease in SERS signals, which can serve as independent feedback mechanisms to signal cargo release from the vesicles [32]. Furthermore, the same group developed SERS active plasmonic vesicles assembled by amphiphilic gold nanorods that can be destructed by both enzymatic degradation and near-infrared photothermal heating, which exhibit a unique combination of optical and structural properties that are of particular interest for theranostic applications [33].

4.3.2 Optical Labeling

Taking advantage of the fingerprinted spectra and high detection sensitivity of SERS nanoprobes, researchers successfully used SERS nanoprobes for the optical labeling of biomolecules through specific intermolecular interaction. By sensing the SERS signals of the nanoprobes, multiplex and ultrasensitive assays at the level of biomolecules, live cells and animals could be realized.

The classic immunoassay was developed for quantitative analysis of molecular targets. Cui's group [35] reported a sensitive immunoassay method for human IgG based on immuno-gold/silver core-shell nanorods SERS nanoprobes modified with goat anti human IgG antibody. The antigen concentration-dependent SERS spectra and dose-response calibration curves were obtained. The detection limit of gold/silver core-shell nanorods based immunoassay reaches 70 fM, which is 10^4 times lower than gold nanorods-based detection. A quick and reproducible tumor marker carcinoembryonic antigen (CEA) analysis method was developed by integrating magnetic beads and SERS nanoprobes [36]. The detection was performed via a two-step process. In the first step, antibody conjugated magnetic beads were added in detection solution containing CEA. Then, the CEA-captured magnetic beads were isolated by a magnetic bar. In the second step, the obtained magnetic beads were further reacted with antibody conjugated SERS nanoprobes. The sandwich immunocomplexes were isolated by using a magnetic bar for further SERS measurements. An LOD of 1–10 pg/mL was obtained. Similarly, Liu et al. devised highly uniform and reproducible SERS nanoprobes by the layer-by-layer assembly of small AgNPs at the surface of SiO_2 particles and applied them for the detection CEA in human serum [37]. The antibody-conjugated $Fe_3O_4@SiO_2$ particles were also used for separating the target molecules in human sera via magnetic force. Taking advantage of the novel structured, highly sensitive SERS nanoprobes, the LOD of this method was lowered to be 0.1 pg mL^{-1}.

The rapid screening and detection of bacterial is an important issue in food safety and medical diagnosis. The multiplex-coded sensitive SERS nanoprobes offer new avenues for rapid screening and analysis. By combining the high sensitivity of SERS nanoprobes with the high specificity of single-domain antibodies (sdAbs), the targeted detection of a single bacterial could be achieved. For instance, the *Salmonella* specific tail spike protein modified silica-encapsulated SERS nanoprobes allowed the detection of a single bacterium using SERS [38]. The selective detection of the multiple drug-resistant bacteria, *Salmonella typhimurium DT104*, was demonstrated by using M3038 monoclonal antibody-conjugated, popcorn-shaped gold nanoprobes, and an LOD of 10 cfu mL^{-1} was achieved [39]. Furthermore, multifunctional popcorn-shaped iron magnetic core-gold plasmonic shell nanoprobes were developed for targeted magnetic separation and enrichment, label-free SERS detection, and the selective photothermal destruction of MDR Salmonella DT104. Besides SERS detection, photothermal-lysis experiment showed that selective and irreparable cellular-damage to MDR Salmonella achieved by using 670 nm light at 1.5 W cm^2 for 10 min. Recently, Chen's group fabricated multifunctional, aldehyde group conjugated graphene oxide wrapped SERS nanoprobes for optical labeling, photothermal ablation of bacteria (Fig. 4.4) [34]. High sensitive Raman imaging of gram-positive (*Staphylococcus aureus*) and gram-negative (*Escherichia coli*) bacteria could be realized and satisfactory photothermal killing efficacy for both bacteria was achieved. The results also demonstrated the correlation among SERS intensity decrease ratio, bacteria survival rate, and the terminal temperature of the nanoprobe-bacteria suspension, showing the possibility to use SERS assay to measure antibacterial response during the photothermal process using this nanoprobe.

SERS nanoprobes had been widely used in live-cell labeling investigations. The SERS technique is suitable for live-cell imaging because strong signals can be produced via low laser powers; Thus, SERS imaging avoids laser-induced injury

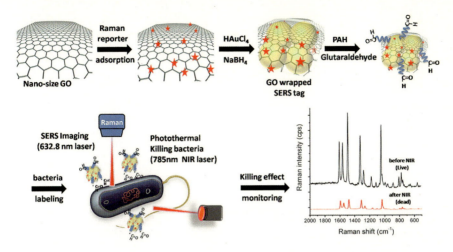

Fig. 4.4 Schematic illustrations of the synthesis of graphene wrapped SERS nanoprobes and the application for optical labeling, photothermal ablation of bacteria, and the monitoring of the killing effect via the thermal-sensitive SERS signal response. Reproduced with permission from Ref. [34]. Copyright 2014, American Chemical Society

of the cells. Furthermore, the excitation laser spot of the Raman microscope can be focused in a micrometer scale. Therefore, the method can provide high-resolution images that reflect the microenvironment in cells. Cancer Marker Detection on the cell membrane was a typical application of SERS nanoprobes in living cells. Walker's group reported the use of antibody-targeted, PEG-coated Au SERS nanoprobes for simultaneous labeling three cell surface markers of interest on malignant B cells from the LY10 lymphoma cell line [40]. The specificity of the nanoprobe' cell labeling was demonstrated on both primary chronic lymphocytic leukemia and LY10 cells using SERS from cell suspensions and confocal Raman mapping. Fluorescence flow cytometry was used to confirm the binding of SERS probes to LY10 over large cell populations, and the nanoprobe' SERS was collected directly from labeled cells using a commercial flow cytometer. Choo and coworkers applied specific antibodies conjugated silica-encapsulated hollow gold nanospheres (SEHGNs) SERS nanoprobes to detect and quantify breast cancer phenotypic markers expressed on cell surfaces [41]. Expression of epidermal growth factor (EGF), ErbB2, and insulin-like growth factor-1 (IGF-1) receptors were assessed in the MDA-MB-468, KPL4 and SK-BR-3 human breast cancer cell lines. SERS imaging were able to test the phenotype of a cancer cell and quantify proteins expressed on the cell surface simultaneously (Fig. 4.5).

SERS nanoprobes had the capability to identify cancer cells in biological samples via specific membrane cancer marker binding. Sha et al. reported detecting circulating breast cancer cells in the blood by using both anti-HER2 antibody-modified SERS nanoprobes and antibody-conjugated, magnetic beads [42]. The magnetic beads could specifically bind to this tumor cell while the SERS

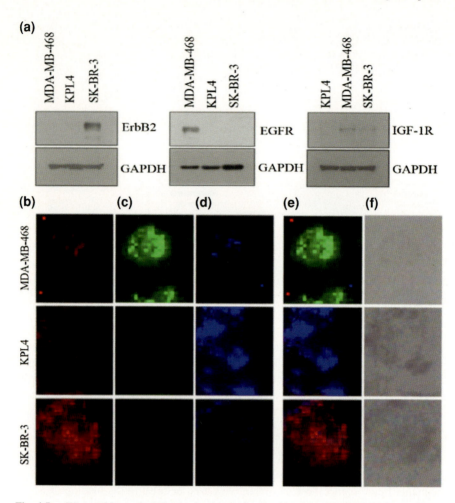

Fig. 4.5 a Western blot analysis for the densitometric quantification of ErbB2, EGFR and IGF-1R markers expressed in MDA468, KPL4 and SK-BR-3 breast cancer cells. The same blot, stripped and reprobed with anti-GAPDH, was used as an internal control. SERS mapping images of corresponding cell lines were measured at (**b**) 1650 cm^{-1} (RBITC), (**c**) 1619 cm^{-1} (MGITC), and (**d**) 1490 cm^{-1} (RuITC). **e** Merged SERS mapping images for three different types of breast cancer cells. **f** Bright field images. Reproduced with permission from Ref. [41]. Copyright 2013 Elsevier

nanoprobes will specifically recognize these breast cancer cells with over expressed HER2 receptors. Thus, the cancer cells could be sensitively detected. SERS nanoprobes with epidermal growth factor (EGF) peptide were applied to identify circulating tumor cells in the peripheral blood of the patients with squamous cell carcinoma of the head and neck, with a range of 1–720 CTCs per milliliter of whole blood [43].

Optical imaging of living subjects is a crucial technique for biomedical research and clinical diagnosis. Up to now, NIR fluorescence, [44, 45] bioluminescence, [46], and photoacoustic tomography, [47] have been extensively used for small-animal models. As a novel tool, Raman imaging has also emerged for optical in vivo imaging analysis that holding two main advantages. First, in vivo multiplex labeling is easily achieved with SERS nanoprobes. Second, the SERS technique can be used with NIR laser excitation, allowing it to share the advantages of fluorescence in vivo imaging in deep tissues. Therefore, SERS imaging shows great potential for real clinical application in the future.

In 2008, Nie and coworkers [25] demonstrated the first use of SERS nanoprobes for in vivo tumor targeting: a single nanoprobe composed of a 60 nm Au NP, crystal violet, and thiol-PEG was approximately 200 times brighter than a NIR QD. When conjugated to tumor-targeting ligands such as single-chain variable fragment antibodies, the SERS nanoprobes could target tumor biomarkers (epidermal growth factor receptors) on human cancer cells and in xenograft tumor models. The SERS spectra obtained by using a 785 nm laser beam on the tumor site had a strong SERS-nanoprobe signature; however, anatomic locations such as the liver only yielded a low background signal.

Later, Gambhir et al. performed a series of work on in vivo SERS imaging techniques. They firstly [48] demonstrated multiplexed in vivo SERS imaging of four types of SERS nanoprobes that had been subcutaneously (s.c.) injected into mice at varying concentrations. Because the SERS nanoprobes had different Raman spectra, the concentration of each could be calculated by using the component analysis method. Furthermore, they demonstrated the ability of Raman spectroscopy to separate the spectral fingerprints of up to ten different s.c. injected nanoprobes and five intravenous (i.v.) injected nanoprobes in liver. They also linearly correlated Raman signals with SERS concentrations after s.c. or i.v. injecting four unique nanoprobes [49]. Despite in vivo multiplex-imaging ability of SERS technique has great potential for medical research, the real application with SERS nanoprobes in endoscopy and intraoperative image guidance of surgical resection was presently limited by long acquisition times and small field of view, and difficult in animal handling with existing Raman spectroscopy instruments. Recently, they presented a unique and dedicated small-animal Raman imaging instrument that enables rapid, high-spatial resolution, spectroscopic imaging over a wide field of view (>6 cm^2), with simplified animal handling (Fig. 4.6). Imaging of SERS nanoprobes in small animals demonstrated that this system could detect multiplexed SERS signals in both superficial and deep tissue locations at least an order of magnitude faster than existing systems without compromising sensitivity [50].

4.3.3 *In situ and Real-Time Monitoring Chemical Reactions*

SERS spectroscopy is an ultrasensitive molecular spectroscopy technique for which detection limits down to the single molecule. Besides, as a vibrational spectroscopy,

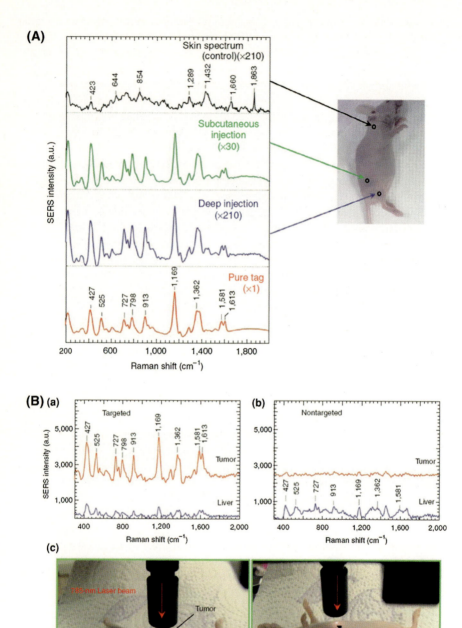

◀ **Fig. 4.6** **A** In vivo SERS spectra obtained from pegylated gold nanoparticles injected into subcutaneous and deep muscular sites in live animals. The injection sites and laser beam positions are indicated by *circles* on the animal. A healthy nude mouse received 50 ml of the SERS nanoparticles tags (1 nM) by subcutaneous (1–2 mm under the skin) or muscular (∼1 cm under the skin) injection. The subcutaneous spectrum was obtained in 3 s, the muscular spectrum in 21 s, and the control spectrum (obtained in an area away from the injection site) also in 21 s. The reference spectrum (*red*) was obtained from the SERS nanoparticles in PBS solution in 0.1 s. The spectral intensities are adjusted for comparison by a factor ($\times 1$, $\times 30$ or $\times 210$) as indicated. The Raman reporter molecule is malachite green, with spectral signatures at 427, 525, 727, 798, 913, 1,169, 1,362, 1,581, and 1,613 cm^{-1}. These features are distinct from the animal skin Raman signals (see the skin spectrum). Excitation wavelength, 785 nm; laser power, 20 mW. **B** In vivo cancer targeting and surface enhanced Raman detection by using ScFv antibody conjugated gold nanoparticles that recognize the tumor biomarker EGFR. **a, b** SERS spectra obtained from the tumor and the liver locations by using targeted (**a**) and nontargeted (**b**) nanoparticles. Two nude mice bearing human head-and-neck squamous cell carcinoma (Tu686) xenograft tumor (3 mm diameter) received 90 µl of ScFv EGFR-conjugated SERS tags or pegylated SERS tags (460 pM). The particles were administered via tail vein single injection. SERS spectra were taken 5 h after injection. **c** Photographs showing a laser beam focusing on the tumor site or on the anatomical location of liver. In vivo SERS spectra were obtained from the tumor site (*red*) and the liver site (*blue*) with 2 s signal integration and at 785 nm excitation. The spectra were background subtracted and shifted for better visualization. The Raman reporter molecule is malachite green, with distinct spectral signatures as labeled in (**a**) and (**b**). Laser power, 20 mW. Reproduced with permission from Ref. [25]. Copyright 2008, Nature Publishing Group

SERS spectroscopy can provide detailed information about the chemical structure of the given molecular target. More importantly, SERS detection can be carried out on micrometer- or submicrometer optically active materials. These features make SERS a powerful tool for the in situ and real-time monitoring chemical reactions.

By using 2 µm hierarchical Ag microspheres with a roughened surface as SERS substrates, Kang et al. demonstrated the monitoring of the chemical reaction of p-nitrothiophenol (pNTP) dimerizing into p,p'-dimercaptoazobenzene. The Plasmon-driven chemical reaction of pNTP could be monitored on one single particle, through which laser wavelength- and power-dependent conversion rates of the reaction were observed (Fig. 4.7a) [51]. Qian et al. reported systematic surface-enhanced Raman studies of two organic chromophores, malachite green (MG) and its isothiocyanate derivative (MGITC). These two dyes have different functional groups for gold NPs surface binding but nearly identical spectroscopic properties. The SERS spectra revealed that the surface structure of MGITC is irreversibly stabilized in its π-conjugated form and is no longer responsive to pH changes. In contrast, the electronic structure of adsorbed MG is still sensitive to pH and can be switched between its localized and delocalized electronic forms [52]. Chemical reactive SERS nanoprobes are helpful for unveiling the mechanism of sodium borohydride ($NaBH_4$) removal of organothiols from AuNPs. When adding $NaBH_4$ to homocysteine (Hcy) or 2-naphthalenethiol (2-NT) labeled SERS nanoprobes, their characteristic Raman peaks disappear. After further spiking 2-NT into Hcy-containing AuNPs that had been washed with $NaBH_4$, the spectrum of 2-NT obtained again. These results indicated that $NaBH_4$ could be used as a hazard-free, general-purpose detergent for AuNP based SERS nanoprobe's recycle and reuse [53].

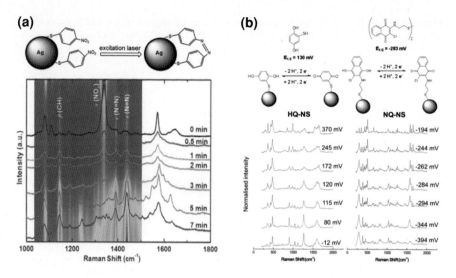

Fig. 4.7 a Time-dependent SERS spectra of p-nitrothiophenol (pNTP) under continuous exposure to a 532 nm laser. The spectra were collected for a single Ag particle, with an integration time of 2 s, and a laser power of 0.5 mW. Reproduced with permission from Ref. [51]. Copyright 2013, Royal Society of Chemistry. **b** Scheme of the local redox potential SERS Nanosensors: structures, electron transfer schemes, and standard reduction potentials and potential-dependent changes in SERS spectra. Reproduced with permission from Ref. [54]. Copyright 2012, American Chemical Society

Redox homeostasis and signaling are critically important in the regulation of cell function, but quantitatively measuring intracellular redox potentials is still challenging. Auchinvole et al. developed a new approach based on the use of SERS nanoprobes, which comprise gold nanoshells and reporter molecules that sense the local redox potential. As shown in Fig. 4.7b, the Raman spectrum of the reporter changes depending on its oxidation state. After the nanoprobes being controllably delivered to the cytoplasm, intracellular potential can be calculated by a simple optical measurement in a reversible, noninvasive manner over a previously unattainable potential range encompassing both superphysiological and physiological oxidative stress [54]. Sharing similar ideas, novel SERS nanoprobes were developed by modifying oxidized cytochrome c (Cyt c) on gold NPs for the sensitive and selective determination superoxide anion radical (O_2^-) in living HeLa and normal human liver cells. On the basis of the differences in the SERS spectra (the height changes of double peaks near 1375 cm^{-1}) between the oxidized and reduced form of Cyt c, this nanoprobe could be used to investigate O_2^- concentration with a detection limit of 1.0×10^{-8} M. Additionally, the selectivity of this nanoprobe was excellent, other reactive oxygen species and biologically relevant species did not influence the detection of O_2^- [55].

4.3.4 Intermolecular Interaction Characterization

A tactfully designed SERS nanoprobe is also suitable for the characterization of protein or nucleic acid interactions inducing conformational changes. By analyzing the vibrational changes occurring at a specific biointerface supported on SERS substrates, interactions between molecules can be in situ monitored and specific targets can be quantitatively detected. Alvarez-Puebla and coworkers did pioneering works in this field. For example, they fabricated silver-coated carbon nanotubes (CNT@Ag) and then modified a monoclonal antibody (mAb) on the surface. When benzoylecgonine, the main cocaine metabolite, met the SERS nanoprobes, it can be captured by mAb. This interaction can be revealed by the increase of Raman intensity of the peak around 693 cm^{-1} (out-of-plane C–H bending) of the nanoprobes (Fig. 4.8a) [56]. This technique was also applied to detect the interaction of the FtsZ protein from *Escherichia coli*, an essential component of the bacterial division machinery, with either a soluble variant of the ZipA protein (that provides membrane tethering to FtsZ) or the bacterial membrane (containing the full-length ZipA naturally incorporated). The engineered silver-coated polystyrene microbeads were used not only to support the bilayers but also to offer a stable support with a high density of SERS hot spots, allowing the detection of ZipA structural changes linked to the binding of FtsZ (Fig. 4.8b)). These changes were different upon incubating the coated beads with FtsZ polymers (GTP form) as compared to oligomers (GDP form) and more pronounced when the plasmonic sensors were coated with natural bacterial membranes [57].

Magnetic hybrid assemblies of Ag and Fe_3O_4 NPs modified with myoglobin (Mb) were able to capture toxic targets (NO_2^-, CN^-, and H_2O_2) and detect them via SERS [58]. Upon binding these targets, the single SERS peak of Mb in the range of 1,340–1,400 cm^{-1} would divide into two peaks. On the basis of characteristic spectral markers, one could quantified these targets with detection limits of 1 nM for nitrite, 0.2 µM for cyanide, and 10 nM for H_2O_2. A novel SERS platform for dopamine (DA) detection was developed based on intermolecular interaction principle. The iron-nitrilotriacetic acid attached silver nanoparticle (Ag–Fe(NTA)) substrate was designed for both DA capture and the SERS enhancement. The Fe-NTA receptors can trap DA adjacent the silver core and the NTA-Fe-DA complex formed provides resonance enhancement with a 632.8 nm laser. DA could be detected in pM level without any pretreatment. The high sensitivity along with the improved selectivity of this sensing approach is a significant step toward molecular diagnosis of Parkinson's disease [59].

The vibrational frequency variations of SERS nanoprobes can also reflect intermolecular interactions. Oncoprotein c-Jun is a member of the bZIP (basic zipper) family of dimeric transcriptional activators whose overexpression has been associated with several human cancers. Guerrini et al. developed SERS-based sensor for the detection of c-Jun to heterodimerize with its native protein partner, c-Fos, and therefore designed a c-Fos peptide receptor chemically modified to incorporate a thiophenol (TP) group at the N-terminal site. As illustrated in

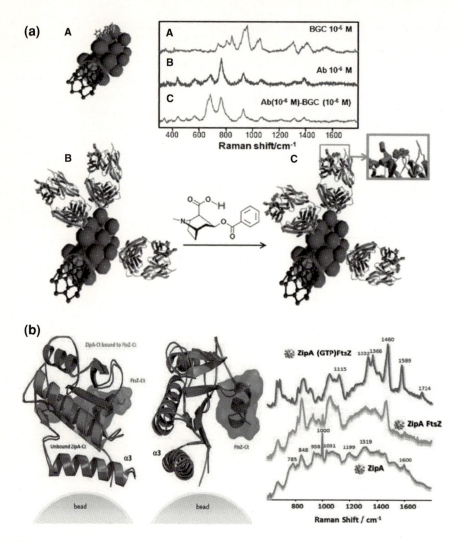

Fig. 4.8 **a** Schematic representation (*green spheres* represent Ag NPs on which the analytes can be retained) and SERS spectra of the direct (**A**) and the label-free specific indirect detection of BCG on CNT@Ag (**B** and **C**). **B** corresponds to the Fab mAb fragments adsorbed on the CNT@Ag substrate whereas (**C**) depicts the complexed mAb–BCG system. Reproduced with permission from Ref. [56]. Copyright 2009, Royal Society of Chemistry. **b** Conformations of sZipA in the absence and in the presence of FtsZ. *Spiral* R-helix; *arrow* β-sheet; *line* random coil. SERS spectra of ZipA on PS@Au@Ag beads, before and after interaction with FtsZ, in the presence and absence of GTP. Samples were illuminated with a 785 nm laser line to avoid damaging the proteins. Reproduced with permission from Ref. [57]. Copyright 2012, American Chemical Society

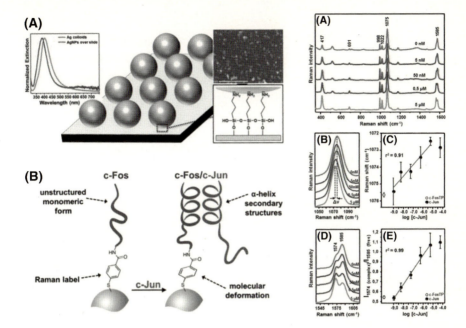

Fig. 4.9 *Left panel* **A** Schematic representation of the SERS substrate (AgNPs over silanized glass slide) including the normalized extinction spectra of AgNPs in solution and deposited on the glass substrate. A representative SEM image of the film is also shown. **B** Outline of the c-Fos/c-Jun dimerization on the metal surface and the resulting deformation of the Raman label structure. c-Fos peptide sequence: MKRRIRRERNKMAAAKCRNRRRELTDTLQAETDQLEDEKSALQ-TEIANLLKEKEKLW. *Right panel* **A** Normalized SERS spectra of c-FosTP anchored to AgNPs over a silanized glass slide upon exposure to variable concentrations of c-Jun in HEPES buffer. **B, D** Details of the 1,000–1,100 and 1,540–1,620 cm^{-1} spectral regions of the SERS spectra illustrated in (A), respectively. **C** Spectral shift of the thiophenol band at ca. 1,075 cm^{-1} as a function of c-Jun concentration (logarithmic scale) in HEPES buffer. **E** Intensity ratio I_{1574} (complex)/I_{1585} (free) as a function of c-Jun concentration (logarithmic scale) in HEPES buffer, with a limit of detection at 5 nM. Reproduced with permission from Ref. [60]. Copyright 2013, American Chemical Society

Fig. 4.9, the binding of two macromolecules induces molecular deformation of TP, the anchors of the c-Fos protein onto the metal substrate, thus the changes of Raman peak position around 1,000–1,100 cm^{-1} and the height ratio of two peaks in 1,540–1,620 cm^{-1} spectral regions. The SERS probes effectively sense the structural rearrangements associated with the c-Fos/c-Jun heterodimerization and can detect c-Jun at nanomolar levels [60]. The vibrational frequency variations of antibody-conjugated SERS nanoprobes were also reported to reflect and quantitatively detect the targeted antigen, as a result of antibody-antigen interaction forces [61].

References

1. Fleischm M, Hendra PJ, McQuilla Aj (1974) Raman-spectra of pyridine adsorbed at a silver electrode. Chem Phys Lett 26(2):163–166
2. Jeanmaire DL, Vanduyne RP (1977) Surface Raman spectroelectrochemistry. 1. heterocyclic, aromatic, and aliphatic-amines adsorbed on anodized silver electrode. J Electroanal Chem 84(1):1–20
3. Campion A, Kambhampati P (1998) Surface-enhanced Raman scattering. Chem Soc Rev 27(4):241–250
4. Kneipp K, Kneipp H, Itzkan I, Dasari RR, Feld MS (1999) Ultrasensitive chemical analysis by Raman spectroscopy. Chem Rev 99(10):2957–2976
5. Stiles PL, Dieringer JA, Shah NC, Van Duyne RP (2008) Surface-enhanced Raman spectroscopy. Annu Rev Anal Chem 1:601–626
6. Lombardi JR, Birke RL (2009) A unified view of surface-enhanced Raman scattering. Acc Chem Res 42(6):734–742
7. Kneipp K, Wang Y, Kneipp H, Perelman LT, Itzkan I, Dasari R, Feld MS (1997) Single molecule detection using surface-enhanced Raman scattering (SERS). Phys Rev Lett 78(9):1667–1670
8. Nie S, Emory SR (1997) Probing Single molecules and single nanoparticles by surface-enhanced Raman scattering. Science 275(5303):1102–1106
9. Jarvis RM, Goodacre R (2008) Characterisation and identification of bacteria using SERS. Chem Soc Rev 37(5):931–936
10. Tripp RA, Dluhy RA, Zhao YP (2008) Novel nanostructures for SERS biosensing. Nano Today 3(3–4):31–37
11. Doering WE, Piotti ME, Natan MJ, Freeman RG (2007) SERS as a foundation for nanoscale, optically detected biological labels. Adv Mater 19(20):3100–3108
12. Banholzer MJ, Millstone JE, Qin L, Mirkin CA (2008) Rationally designed nanostructures for surface-enhanced Raman spectroscopy. Chem Soc Rev 37(5):885–897
13. Wang Y, Yan B, Chen L (2013) SERS tags: novel optical nanoprobes for bioanalysis. Chem Rev 113(3):1391–1428
14. Kneipp J, Kneipp H, Wittig B, Kneipp K (2010) Novel optical nanosensors for probing and imaging live cells. Nanomedicine 6(2):214–226
15. Dinish US, Balasundaram G, Chang YT, Olivo M (2014) Actively targeted in vivo multiplex detection of intrinsic cancer biomarkers using biocompatible SERS nanotags. Sci Rep. doi:10.1038/srep04075
16. Xie J, Zhang Q, Lee JY, Wang DI (2008) The synthesis of SERS-active gold nanoflower tags for in vivo applications. ACS Nano 2(12):2473–2480
17. Pinkhasova P, Yang L, Zhang Y, Sukhishvili S, Du H (2012) Differential SERS activity of gold and silver nanostructures enabled by adsorbed poly(vinylpyrrolidone). Langmuir 28(5):2529–2535
18. Tam NC, McVeigh PZ, Macdonald TD, Farhadi A, Wilson BC, Zheng G (2012) Porphyrin-lipid stabilized gold nanoparticles for surface enhanced Raman scattering based imaging. Bioconjug Chem 23(9):1726–1730
19. Tam NC, Scott BM, Voicu D, Wilson BC, Zheng G (2010) Facile synthesis of Raman active phospholipid gold nanoparticles. Bioconjug Chem 21(12):2178–2182
20. Liu X, Knauer M, Ivleva NP, Niessner R, Haisch C (2010) Synthesis of core-shell surface-enhanced Raman tags for bioimaging. Anal Chem 82(1):441–446
21. Küstner B, Gellner M, Schütz M, Schöppler F, Marx A, Ströbel P, Adam P, Schmuck C, Schlücker S (2009) SERS labels for red laser excitation: silica-encapsulated SAMs on tunable gold/silver nanoshells. Angew Chem Int Ed 48(11):1950–1953
22. Schlücker S (2009) SERS microscopy: nanoparticle probes and biomedical applications. ChemPhysChem 10(9–10):1344–1354

23. Li ZY, Xia Y (2010) Metal nanoparticles with gain toward single-molecule detection by surface-enhanced Raman scattering. Nano Lett 10(1):243–249

24. Doering WE, Nie S (2003) Spectroscopic tags using dye-embedded nanoparticles and surface-enhanced Raman scattering. Anal Chem 75(22):6171–6176

25. Qian X, Peng XH, Ansari DO, Yin-Goen Q, Chen GZ, Shin DM, Yang L, Young AN, Wang MD, Nie S (2008) In vivo tumor targeting and spectroscopic detection with surface-enhanced Raman nanoparticle tags. Nat Biotechnol 26(1):83–90

26. Lou T, Wang Y, Li J, Peng H, Xiong H, Chen L (2011) Rapid detection of melamine with 4-mercaptopyridine-modified gold nanoparticles by surface-enhanced Raman scattering. Anal Bioanal Chem 401(1):333–338

27. Liu TY, Tsai KT, Wang HH, Chen Y, Chen YH, Chao YC, Chang HH, Lin CH, Wang JK, Wang YL (2011) Functionalized arrays of Raman-enhancing nanoparticles for capture and culture-free analysis of bacteria in human blood. Nat Commun 2:538

28. Yin J, Wu T, Song JB, Zhang Q, Liu SY, Xu R, Duan HW (2011) SERS-active nanoparticles for sensitive and selective detection of cadmium ion (Cd^{2+}). Chem Mater 23(21):4756–4764

29. Ye Y, Liu H, Yang L, Liu J (2012) Sensitive and selective SERS probe for trivalent chromium detection using citrate attached gold nanoparticles. Nanoscale 4(20):6442–6448

30. Chen L, Fu X, Li J (2013) Ultrasensitive surface-enhanced Raman scattering detection of trypsin based on anti-aggregation of 4-mercaptopyridine-functionalized silver nanoparticles: an optical sensing platform toward proteases. Nanoscale 5(13):5905–5911

31. Wang X, Chen L, Fu X, Ding Y (2013) Highly sensitive surface-enhanced Raman scattering sensing of heparin based on anti-aggregation of functionalized silver nanoparticles. ACS Appl Mater Interfaces 23;5(21):11059–1165

32. Song J, Zhou J, Duan H (2012) Self-assembled plasmonic vesicles of SERS-encoded amphiphilic gold nanoparticles for cancer cell targeting and traceable intracellular drug delivery. J Am Chem Soc 134(32):13458–13469

33. Song JB, Pu L, Zhou JJ, Duan B, Duan HW (2013) Biodegradable theranostic plasmonic vesicles of amphiphilic gold nanorods. ACS Nano 7(11):9947–9960

34. Lin D, Qin T, Wang Y, Sun X, Chen L (2014) Graphene oxide wrapped SERS tags: multifunctional platforms toward optical labeling, photothermal ablation of bacteria, and the monitoring of killing effect. ACS Appl Mater Interfaces 6(2):1320–1329

35. Wu L, Wang Z, Zong S, Huang Z, Zhang P, Cui Y (2012) A SERS-based immunoassay with highly increased sensitivity using gold/silver core-shell nanorods. Biosens Bioelectron 38(1):94–99

36. Chon H, Lee S, Son SW, Oh CH, Choo J (2009) Highly sensitive immunoassay of lung cancer marker carcinoembryonic antigen using surface-enhanced Raman scattering of hollow gold nanospheres. Anal Chem 81(8):3029–3034

37. Liu R, Liu B, Guan G, Jiang C, Zhang Z (2012) Multilayered shell SERS nanotags with a highly uniform single-particle Raman readout for ultrasensitive immunoassays. Chem Commun 48(75):9421–9423

38. Tay LL, Huang PJ, Tanha J, Ryan S, Wu X, Hulse J, Chau LK (2012) Silica encapsulated SERS nanoprobe conjugated to the bacteriophage tailspike protein for targeted detection of Salmonella. Chem Commun 48(7):1024–1026

39. Khan SA, Singh AK, Senapati D, Fan Z, Ray PC (2011) Targeted highly sensitive detection of multi-drug resistant Salmonella DT104 using gold nanoparticles. Chem Commun 47(33):9444–9446

40. Maclaughlin CM, Mullaithilaga N, Yang G, Ip SY, Wang C, Walker GC (2013) Surface-enhanced Raman scattering dye-labeled au nanoparticles for triplexed detection of leukemia and lymphoma cells and SERS flow cytometry. Langmuir 29(6):1908–1919

41. Lee S, Chon H, Lee J, Ko J, Chung BH, Lim DW, Choo J (2013) Rapid and sensitive phenotypic marker detection on breast cancer cells using surface-enhanced Raman scattering (SERS) imaging. Biosens Bioelectron 51:238–243

42. Sha MY, Xu H, Natan MJ, Cromer R (2008) Surface-enhanced Raman scattering tags for rapid and homogeneous detection of circulating tumor cells in the presence of human whole blood. J Am Chem Soc 130(51):17214–17215
43. Wang X, Qian X, Beitler JJ, Chen ZG, Khuri FR, Lewis MM, Shin HJ, Nie S, Shin DM (2011) Detection of circulating tumor cells in human peripheral blood using surface-enhanced Raman scattering nanoparticles. Cancer Res 71(5):1526–1532
44. Wang Y, Ye C, Wu L, Hu Y (2010) Synthesis and characterization of self-assembled CdHgTe/gelatin nanospheres as stable near infrared fluorescent probes in vivo. J Pharm Biomed Anal 53(3):235–242
45. Chen H, Wang Y, Xu J, Ji J, Zhang J, Hu Y, Gu Y (2008) Non-invasive near infrared fluorescence imaging of CdHgTe quantum dots in mouse model. J Fluoresc 18(5):801–811
46. Dothager RS, Flentie K, Moss B, Pan MH, Kesarwala A, Piwnica-Worms D (2009) Advances in bioluminescence imaging of live animal models. Curr Opin Biotechnol 20(1):45–53
47. Li C, Wang LV (2009) Photoacoustic tomography and sensing in biomedicine. Phys Med Biol 54(19):R59–R97
48. Keren S, Zavaleta C, Cheng Z, de la Zerda A, Gheysens O, Gambhir SS (2008) Noninvasive molecular imaging of small living subjects using Raman spectroscopy. Proc Natl Acad Sci USA 105(15):5844–5849
49. Zavaleta CL, Smith BR, Walton I, Doering W, Davis G, Shojaei B, Natan MJ, Gambhir SS (2009) Multiplexed imaging of surface enhanced Raman scattering nanotags in living mice using noninvasive Raman spectroscopy. Proc Natl Acad Sci USA 106(32):13511–13516
50. Bohndiek SE, Wagadarikar A, Zavaleta CL, Van de Sompel D, Garai E, Jokerst JV, Yazdanfar S, Gambhir SS (2013) A small animal Raman instrument for rapid, wide-area, spectroscopic imaging. Proc Natl Acad Sci USA 110(30):12408–12413
51. Kang L, Xu P, Zhang B, Tsai H, Han X, Wang HL (2013) Laser wavelength- and power-dependent plasmon-driven chemical reactions monitored using single particle surface enhanced Raman spectroscopy. Chem Commun 49(33):3389–3391
52. Qian X, Emory SR, Nie S (2012) Anchoring molecular chromophores to colloidal gold nanocrystals: surface-enhanced Raman evidence for strong electronic coupling and irreversible structural locking. J Am Chem Soc 134(4):2000–2003
53. Ansar SM, Ameer FS, Hu W, Zou S, Pittman CU Jr, Zhang D (2013) Removal of molecular adsorbates on gold nanoparticles using sodium borohydride in water. Nano Lett 13(3):1226–1229
54. Auchinvole CA, Richardson P, McGuinnes C, Mallikarjun V, Donaldson K, McNab H, Campbell CJ (2012) Monitoring intracellular redox potential changes using SERS nanosensors. ACS Nano 6(1):888–896
55. Qu LL, Li DW, Qin LX, Mu J, Fossey JS, Long YT (2013) Selective and sensitive detection of intracellular O using Au NPs/cytochrome c as SERS nanosensors. Anal Chem 85(20):9549–9555
56. Sanles-Sobrido M, Rodriguez-Lorenzo L, Lorenzo-Abalde S, Gonzalez-Fernandez A, Correa-Duarte MA, Alvarez-Puebla RA, Liz-Marzan LM (2009) Label-free SERS detection of relevant bioanalytes on silver-coated carbon nanotubes: the case of cocaine. Nanoscale 1(1):153–158
57. Ahijado-Guzman R, Gomez-Puertas P, Alvarez-Puebla RA, Rivas G, Liz-Marzan LM (2012) Surface-enhanced Raman scattering-based detection of the interactions between the essential cell division ftsz protein and bacterial membrane elements. ACS Nano 6(8):7514–7520
58. Han XX, Schmidt AM, Marten G, Fischer A, Weidinger IM, Hildebrandt P (2013) Magnetic silver hybrid nanoparticles for surface-enhanced resonance Raman spectroscopic detection and decontamination of small toxic molecules. ACS Nano 7(4):3212–3220
59. Kaya M, Volkan M (2012) New approach for the surface enhanced resonance Raman scattering (SERRS) detection of dopamine at picomolar levels in the presence of ascorbic acid. Anal Chem 84(18):7729–7735

60. Guerrini L, Pazos E, Penas C, Vazquez ME, Mascarenas JL, Alvarez-Puebla RA (2013) Highly sensitive SERS quantification of the oncogenic protein c-Jun in cellular extracts. J Am Chem Soc 135(28):10314–10317
61. Kho KW, Dinish US, Kumar A, Olivo M (2012) Frequency shifts in SERS for biosensing. ACS Nano 6(6):4892–4902

Chapter 5
Challenges and Perspectives of Optical Nanoprobes

Abstract In recent years, the design of optical nanoprobes has achieved dramatic advances and certain nanosensors have been successfully applied to biological analysis and imaging. Still, challenges of optical nanoprobes also remain, such as increasing sensing performance for real environmental and clinical samples, the design and application of multifunctional nanoplatforms, and biocompatibility research.

Keywords Optical nanoprobes · Real sample detection · Multifunctional nanoplatform · Biocompatibility

Despite various kinds of optical nanoprobes have been developed, it should be noted that most academic researches stayed on the level of exploring novel optical features of nanomaterials and developing proof-of-concept sensors that only fit for detection in "pure" sensing systems. Up to now, no nanoprobe is really applicable to monitor the chemical parameter under real-world conditions, for example in the blood stream, in seawater, or in a reaction vessel [1]. This problem is partially attributed to the active nature of nanomaterials. For example, monodispersed gold or silver nanoparticles (NPs)-based colorimetric probes themselves are likely to aggregate in ionic buffers and lose sensing ability. The fluorescence of quantum dots (QDs) is sensitive to a variety of substances in environment or in physiological matrix such as H^+, Fe^{3+}, and H_2O_2 [2, 3]. Therefore, the selectivity for real sample detection is very hard to achieve when using QDs-based nanoprobes. Gold NPs tend to adsorb endogenous macromolecules due to the large specific surface area, which result in inactive surface and decrease the surface enhanced Raman scattering (SERS) ability [4]. This is a major obstacle for the application of SERS sensing nanoprobes in vivo. In the future, researchers should face and resolve these questions of reality and tried their best to deliver numerous kinds of optical nanoprobes from lab to real and urgent application scenes and markets.

Developing multifunctional nanoplatforms combining the benefits of multiple imaging and diagnosis capacities will be another important and practical research field [5]. These complex probes exhibit much extended sensing capability and have great potential to conquer challenges such as delineating tumor margins during surgery and tracking in vivo distribution behavior of nanomedicine. Two factors should be considered in this issue. The first factor is rational design strategies to preserve the optical properties of multiple kinds of nanomaterials during constructing complex nanostructures composed of multiple targeting, separation, imaging, and therapeutic modules. For example, unique triple-modality magnetic resonance imaging–photoacoustic imaging–Raman imaging NPs were reported to accurately delineate the margins of brain tumors in living mice both preoperatively and intraoperatively [6]. Multifunctional gold nanorods with ultrahigh stability and tunability were designed for in vivo fluorescence imaging, SERS detection, and photodynamic therapy [7]. The second factor to be noted is how to maintain a relative small size of the resultant multicomponent nanostructures. In biomedical analysis, if the nanoprobes are too large compared with targeting biological molecules, any inhomogeneous distributions of the probes superimposed on the biological samples will reduce the accuracy of bioimaging and analysis. Thus, size controlling of the probes is important to accurately reveal the location and number of the targets in sensing images. On one hand, small-sized NPs (sub 10 nm) such as gold/silver nanoclusters, [8] carbon dots, [9, 10] silicon dots, [11–13], graphene QDs [14], and small dye-encapsulated SERS tags [15, 16] were developed and used as raw materials. On the other hand, delicate core-shell and yolk-shell architectures [17, 18] were skillfully designed to reduce the size of the complex multifunctional probes.

Biocompatibility of nanoprobes is an important issue in bioanalysis, which mainly depends on the toxicity of corresponding nanomaterials. In recent years, the biological effect of various types of continuously emerging nanomaterials has drawn much concern [19]. Conclusive evidences indicated that many NPs for building up optical probes were of potential toxicity for living systems. For example, QDs could induce adverse effect to cells due to slow oxidation process induced free cadmium ions release, and the creation of reactive oxygen species that cause damage to nucleic acids and enzymes [20]. Gold nanoclusters showed size-dependent toxicity, small clusters with size less than 2 nm was much toxic than 15 nm gold NPs [21]. Therefore, surface stabilizing and coating materials and protocols should be performed to reduce the toxicity of nanomaterials, and the as-prepared probes should be subject to biocompatibility test before applying for in vivo investigations.

Additionally, by taking advantage of the multiplex information provided by optical nanoprobes, novel instruments such as flow cytometers for cell sorting [22] and miniaturized sensor arrays, microfluidic chips [23] for clinical analysis can be further developed. Future environmental sensors should be envisioned to be integrated into mobile phones, computers, or aerial drones for real-time monitoring and remote data transmission.

References

1. Wolfbeis OS (2013) Probes, sensors, and labels: why is real progress slow? Angew Chem Int Ed 52(38):9864–9865
2. Wang YQ, Ye C, Zhu ZH, Hu YZ (2008) Cadmium telluride quantum dots as pH-sensitive probes for tiopronin determination. Anal Chim Acta 610(1):50–56
3. Hu M, Tian J, Lu HT, Weng LX, Wang LH (2010) H_2O_2-sensitive quantum dots for the label-free detection of glucose. Talanta 82(3):997–1002
4. Wang Y, Yan B, Chen L (2013) SERS tags: novel optical nanoprobes for bioanalysis. Chem Rev 113(3):1391–1428
5. Kim J, Piao Y, Hyeon T (2009) Multifunctional nanostructured materials for multimodal imaging, and simultaneous imaging and therapy. Chem Soc Rev 38(2):372–390
6. Kircher MF, de la Zerda A, Jokerst JV, Zavaleta CL, Kempen PJ, Mittra E, Pitter K, Huang RM, Campos C, Habte F, Sinclair R, Brennan CW, Mellinghoff IK, Holland EC, Gambhir SS (2012) A brain tumor molecular imaging strategy using a new triple-modality MRI-photoacoustic-Raman nanoparticle. Nat Med 18(5):829–U235
7. Zhang Y, Qian J, Wang D, Wang Y, He S (2013) Multifunctional gold nanorods with ultrahigh stability and tunability for in vivo fluorescence imaging, SERS detection, and photodynamic therapy. Angew Chem Int Ed 52(4):1148–1151
8. Li N, Chen Y, Zhang YM, Yang Y, Su Y, Chen JT, Liu Y (2014) Polysaccharide-gold nanocluster supramolecular conjugates as a versatile platform for the targeted delivery of anticancer drugs. Sci Rep 4:4164
9. Cao L, Yang ST, Wang X, Luo PG, Liu JH, Sahu S, Liu Y, Sun YP (2012) Competitive performance of carbon quantum dots in optical bioimaging. Theranostics 2(3):295–301
10. Zhu S, Meng Q, Wang L, Zhang J, Song Y, Jin H, Zhang K, Sun H, Wang H, Yang B (2013) Highly photoluminescent carbon dots for multicolor patterning, sensors, and bioimaging. Angew Chem Int Ed 52(14):3953–3957
11. Liu J, Erogbogbo F, Yong KT, Ye L, Hu R, Chen H, Hu Y, Yang Y, Yang J, Roy I, Karker NA, Swihart MT, Prasad PN (2013) Assessing clinical prospects of silicon quantum dots: studies in mice and monkeys. ACS Nano 7(8):7303–7310
12. He Y, Zhong Y, Peng F, Wei X, Su Y, Lu Y, Su S, Gu W, Liao L, Lee ST (2011) One-pot microwave synthesis of water-dispersible, ultraphoto- and pH-stable, and highly fluorescent silicon quantum dots. J Am Chem Soc 133(36):14192–14195
13. Peng F, Su Y, Zhong Y, Fan C, Lee ST, He Y (2014) Silicon nanomaterials platform for bioimaging, biosensing, and cancer therapy. Acc Chem Res 47(2):612–623
14. Shen J, Zhu Y, Yang X, Li C (2012) Graphene quantum dots: emergent nanolights for bioimaging, sensors, catalysis and photovoltaic devices. Chem Commun 48(31):3686–3699
15. Zhang P, Guo Y (2009) Surface-enhanced Raman scattering inside metal nanoshells. J Am Chem Soc 131(11):3808–3809
16. Li W, Guo Y, Zhang P (2010) A general strategy to prepare TiO_2-core gold-shell nanoparticles as SERS-tags. J Phys Chem C 114(16):7263–7268
17. Guerrero-Martinez A, Perez-Juste J, Liz-Marzan LM (2010) Recent progress on silica coating of nanoparticles and related nanomaterials. Adv Mater 22(11):1182–1195
18. Fan W, Shen B, Bu W, Chen F, Zhao K, Zhang S, Zhou L, Peng W, Xiao Q, Xing H, Liu J, Ni D, He Q, Shi J (2013) Rattle-structured multifunctional nanotheranostics for synergetic chemo-/radiotherapy and simultaneous magnetic/luminescent dual-mode imaging. J Am Chem Soc 135(17):6494–6503
19. Zhang Y, Bai Y, Jia J, Gao N, Li Y, Zhang R, Jiang G, Yan B (2014) Perturbation of physiological systems by nanoparticles. Chem Soc Rev 43(10):3762–3809
20. Wang Y, Chen L (2011) Quantum dots, lighting up the research and development of nanomedicine. Nanomedicine 7(4):385–402
21. Pan Y, Neuss S, Leifert A, Fischler M, Wen F, Simon U, Schmid G, Brandau W, Jahnen-Dechent W (2007) Size-dependent cytotoxicity of gold nanoparticles. Small 3(11):1941–1949

22. Goddard G, Brown LO, Habbersett R, Brady CI, Martin JC, Graves SW, Freyer JP, Doorn SK (2010) High-resolution spectral analysis of individual SERS-active nanoparticles in flow. J Am Chem Soc 132(17):6081–6090
23. Cecchini MP, Hong J, Lim C, Choo J, Albrecht T, Demello AJ, Edel JB (2011) Ultrafast surface enhanced resonance Raman scattering detection in droplet-based microfluidic systems. Anal Chem 83(8):3076–3081